SpringerWienNewYork

Gerhard Nierhaus

Algorithmic Composition

Paradigms of Automated Music Generation

SpringerWienNewYork

Gerhard Nierhaus
Institute of Electronic Music and Acoustics (IEM)
University of Music and Dramatic Arts, Graz, Austria

Printed in Germany

SpringerWienNewYork is part of Springer Science + Business Media
springer.at

Printed on acid-free and chlorine-free bleached paper

With 173 figures and 20 tables

ISBN 978-3-211-99915-8

e-ISBN 978-3-211-75540-2

Acknowledgements

The writing and editing of this book presents a large effort on many people and institutions who I want to express my gratitude to. Without their help, this work would not have been completed.

First, I wish to thank the Austrian Science Fund (FWF), the Styrian Provincial Government's Science and Research Department, the City of Graz Department of Cultural Affairs and the Society for the Promotion of Electronic Music and Acoustics (GesFEMA) for their generous financial support.

My sincere thanks to Ilona Leimisch, who translated the book into English, for her efforts and patience during our collaboration.

In particular I am grateful for the help of the former head of the Institute of Electronic Music and Acoustics Graz (IEM) at the University of Music and Dramatic Arts Graz, Robert Höldrich, who contributed valuable advice. I also wish to thank several people who have provided their kind assistance with proofreading. These include the present head of the IEM, Gerhard Eckel, as well as Alberto de Campo, Markus Noisternig, Alois Sontacchi and Camilla Leimisch, who also helped editing the book.

Many institutions and publishers have given generous support concerning the illustrative material. I would particularly like to name and thank (in alphabetical order) A-R editions, Badische Landesbibliothek, Carl Hanser Verlag, Casa Editrice Leo S. Olschki, Dover Publications, IBM, Institute for Advanced Study, Los Alamos National Laboratories, Massachusetts Institute of Technology, Moravian Library, Naval Historical Center, Pearson Education, Pendragon Press, Princeton University Library, Research Library in Olomouc, Technisches Museum Wien and Wolfram Media for providing interesting material free of charge.

In addition to the abovementioned, I acknowledge with appreciation the assistance of some significant scientists for making available pictures and other material free of charge that were of great value for the book. These are (in alphabetical order) Moray Allan, Bernard Bel, Wei Chai, Noam Chomsky, David Cope, Roger Luke DuBois, Sidney Fels, David Goldberg, Ricardo Gudwin, Douglas Hofstadter, John H. Holland, Philip Johnson-Laird, the recently deceased Edward Lorenz, Benoit Mandelbrot, Jonatas Manzolli, Herbert Matis, Artemis Moroni, Craig Nevill-

Manning, Somnuk Phon-Amnuaisuk, Alan Smaill, Mark Steedman, Andrew Tuson, Geraint Wiggins, Stephen Wolfram and Ferdinando von Zuben.

Lastly, I want to thank Springer-Verlag and especially Stephen Soehnlen, whose generous contribution of time and expertise were of invaluable importance for the development of the book.

Contents

Chapter 1
Introduction

Algorithmic composition – composing by means of formalizable methods – has a century old tradition not only in occidental music history. Guido of Arezzo who, with his "Micrologus de disciplina artis musicae" created a groundbreaking work of music theory as early as 1025, is today mainly known for inventing 'solfeggio' as well as due to his essential contributions to the development of musical notation. Less known is Arezzo's work in the field of algorithmic composition where he invented a method for the automatic conversion of text into melodic phrases.

The inclusion of rules and general frameworks into the creative process of composition, however, becomes apparent not only in exemplary systems which generate musical structures "at the push of a button"; it is also expressed in a number of works throughout music history. Well-known examples such as Bach's"The Art of Fugue" or Schönberg's twelve-tone technique immediately come to mind which often leads to placing the development of formal complexity exclusively into the context of European, or occidental, art music. One reason for the nearsightedness of this approach which excludes numerous "non-European" music traditions from investigation can definitely be found in the over-emphasis of European art music in musical education; the term "non-European" incidentally points out this problem very clearly – would you, for example, refer to the sonata as a "non-Australian" music genre?

This book, however, does not deal with the treatment of formalizable aspects in single works or particular musical traditions, but attempts to provide an overview of prominent procedures of algorithmic composition in a pragmatic way. Though some passages in this work nevertheless refer to the working methods of specific composers or formal principles in certain music genres, these are only mentioned as examples and do not present an exhaustive enumeration. So, for example, the examination of algorithmic principles in the works of composers such as Xenakis or König alone goes far beyond the scope of this book and would furthermore represent an analytical musicologist approach not intended here. Also, the comprehensive and rapidly developing field of mathematical music theory,[1] which opens up innovative approaches not only for computer assisted analysis, cannot be treated in this work.

[1] So, as a prime example the comprehensive work of Guerino Mazzola [3].

If one now poses the legitimate question about which musical approaches actually may be subsumed under the term "algorithmic composition," an answer can be found in the investigation of some general definitions of "algorithm." Etymologically, this term may be derived from the Greek word "arithmos" (number) and also from the name of the Persian mathematician Abu Jafar Muhammad ibn Musa al-Khwarizmi[2] in two different ways. Around AD 820, al-Khwarizmi wrote a treatise on the calculation with Indian numerals, which was translated into Latin around 1120 AD as "Algorismi de numero Indorum." In this translation, the author was given the Latinized name "Algorismus." In "Carmen de Algorismo," a mathematical treatise in rhymes by French scholar Alexander de Villa Dei from AD 1220, calculating in the new numeral systems is also referred to with the term "Algorismus."[3] Later, it was Grecized and became "Algorithmus," being used as a general indication for a controlled procedure. Let us now turn to some definitions of the term, e.g.:

- A set of mathematical instructions that must be followed in a fixed order, and that, especially if given to a computer, will help to calculate an answer to a mathematical problem [1].
- A systematic procedure that produces – in a finite number of steps – the answer to a question or the solution of a problem [2].
- [...] (especially computing) a set of rules that must be followed when solving a particular problem [4].

For a 'classical algorithm,' in general also a number of conditions may apply, e.g. it must *terminate*, i.e. produce solutions after running for a finite time of steps, or be *deterministic*, i.e. with a particular input, it will always produce the same correct output. However, some classes of algorithms and systems of algorithmic composition may produce output indefinitely, or involve probability-based decisions in the problem solving process. In many cases, for example, in control algorithms within operating systems, terminating behavior can be observed frequently, but the fulfillment of this condition is nevertheless not decidedly desired. Similarly, an algorithm for the generation of musical structure will not terminate, if, for example, new material is to be made continuously audible for the arrangement of a sound installation. The involvement of probability allows for *stochastic* algorithms to produce different results with the same initial values – this approach can be found in most systems designed for algorithmic composition. Disregarding this special classification, an algorithm may be, based on the abovementioned definitions, very generally described as a formalizable and abstracting procedure which – applied to the generation of musical structure – determines the field of application of algorithmic composition.

Accordingly, models for generating musical structure may be obtained from nearly every scientific discipline. Apart from that, interesting musical results may also be reached through simple but innovative compositional strategies or an ap-

[2] "Mohammed the father of Jafar and the son of Musa, the Khwarizmian," also referred to as "al Khowarizmi," lived ca. 780–850.

[3] In Latin also "alchorismus" and "algoarismus"; in Old French "algorisme" and "argorisme"; in Middle English "augrim" and "augrym."

propriate mapping of data onto musical parameters. Necessary restrictions for the framework of this book also result from these comprehensive application possibilities: Procedures are described here that on the one hand are very well suited to the generation of musical structure, and on the other hand each represent a class of algorithms that can process or generate musical information in a specific way. Herein, the scope ranges from generative grammars that process and produce musical material in concise formalisms, to cellular automata in which a few simple initial rules lead to a complex behavior.

For the musical applications of the specific classes of algorithms, only examples are selected that – in the stricter sense of algorithmic composition – treat the generation of musical structure on a symbolical level. This means that in these cases the outputs mostly represent control data or note values, whose acoustical realization – be it with conventional instruments or synthesis processes – is no longer a task of the generating algorithm. This delimitation may, of course, also be seen as a restriction because the structure and sonic detail of a musical event are in general closely connected factors.

Within the respective procedures, music-specific knowledge is made available in different ways: Rules for the generation of musical structure are either formulated or generated through the analysis of a corpus. The type of the examined procedures and their strategy of knowledge representation promote different approaches to the production of structure that are either applied for the generation of 'style imitations' – the generation of musical material according to a given notion of a musical style – or as a means of implementing compositional strategies for the creation of a new piece of art, which may be called 'genuine composition.'

The single chapters

After a historic overview, the following chapters each present a specific class of algorithm in the context of algorithmic composition, first providing a general introduction to their development and their theoretical basics, and then describing different musical applications. At the end of each chapter, strengths and weaknesses as well as possible aesthetical implications resulting from the application of the treated approaches are outlined. Since the discussed algorithm classes each enable a specific approach to musical structure generation, one could in this context also speak of different "paradigms" of algorithmic composition.

In the chapter **"Historical Development of Algorithmic Procedures,"** innovations are described, beginning with the first documented use of counting to the invention of the computer as it is today, which form the basis for the different applications of algorithmic composition. Technical improvements do not present singular events, but are produced in the context of a sociological, sociopolitical and philosophical framework. The automatization of procedures is preceded by necessary processes of abstraction that may be traced back to a number of historical developments. Here, the scope ranges from the creation of numeral and writing systems, the formalization of thinking processes up to attempts to automatize the establishment of truth in systems of "universal validity." Primary principles of automated information processing can already be found in the 13th century. Through the works of Charles Babbage and Ada Lovelace, functioning concepts of modern computer architecture were developed already hundred years before the creation of the "first" digital computers by Konrad Zuse and Howard Hathaway Aiken. The fact that world view and scientific advancement are interrelated factors of human societies can be seen in aspects such as the invention of the number zero or the philosophical implications of probability theory. The history of algorithmic composition has its beginning shortly after the turn of the first millennium with a system developed by Guido of Arezzo enabling the generation of melodic material from texts, spans over the application of algorithmic principles in the developing complex polyphony and is also found in the "composition machines" of Athanasius Kircher in the Baroque period. Furthermore, first applications of algorithms for compositional tasks can be found in the popular "musical dice game" in the 18th century. Finally, on August 09, 1956 the "Illiac Suite," the first computer-generated composition, had its world premiere at the University of Illinois. In regard to the history of algorithmic composition, this date, however, marks only the beginning of a number of forthcoming fascinating developments which involve technical advancements as well as the investigation of new scientific disciplines.

Markov models, which were originally developed in the context of language processing, are a well-established paradigm of algorithmic composition. For the generation of structure, a sequence of states and also their transition probabilities are extracted by analysis of a corpus. The number of preceding states that are applied for the calculation of the new state is determined by the order of the Markov model – higher orders produce sequences that are more and more similar to the corpus, but may also have disadvantageous effects in musical structure generation. Hidden Markov models present an interesting variant, since observable events enable inferences on hidden underlying state changes. Markov models are for the most part employed in the field of style imitation, but also, for example by Hiller and Xenakis, for applications of genuine composition.

With **generative grammars**, a powerful formalism for the generation of musical structure establishes itself in algorithmic composition. Developed in the context of linguistics as well, they allow for the production of structures that are context-sensitive over long passages. In the field of algorithmic composition, generative grammar is above all used according to the hierarchy created by the American linguist Noam Chomsky. This approach distinguishes grammar types of different expressiveness – higher types are easier to manage, but inferior to lower types regarding their generative capacity. Grammatical inference enables the automatic generation of rewriting rules from a corpus – in contrast to Markov models, different context depths of a musical material may be comprehended in this case at the same time.

Transition networks, which are represented in a graph, may be applied to a broad range of tasks in algorithmic composition. David Cope's well-known program "EMI" uses this formalism for representing and processing musical information. "EMI" generates style imitations after having analyzed a sufficient number of compositions of a particular genre. Cope's system applies a complex strategy in the recombination of musical segments, considering their formal significance on different hierarchical levels. Transition networks exist in different forms, such as Petri nets, and allow for an adaptation to distinct musical tasks.

In the 1980s, **chaos and self-similarity** became catchwords in a broad discussion that reached far beyond scientific communities through the much-cited "butterfly effect," a wide range of fascinating graphical realizations of fractals and other aspects of the heterogeneous field of chaos theory. For algorithmic composition, these innovations opened interesting possibilities for generating complex musical structures whose temporal developments react highly sensitively to modifications in the initial configurations. Lindenmayer systems were originally developed for the simulation of the growth process of plants and are well suited to the simulation of self-similar processes. The different variants of Lindenmayer systems enable the implementation of a wide range of compositional concepts for the field of algorithmic composition.

Genetic algorithms and genetic programming are probability-based search techniques created on the model of Darwin's theory of evolution. Problems are solved by applying quasi-biological procedures in a virtual biological environment. Genetic operations such as crossover and mutation produce new individuals in every generation that are evaluated according to a quality criterion, which has to be met. The specific behavior of a genetic algorithm promotes primarily compositional approaches which create musical material as a time-flow, i.e. a process of continuous change. An interesting application of genetic algorithms can be found in some works that make the interactions of "individuals" in an artificial habitat audible in an innovative way.

Cellular automata were first created in the 1940s and gained great popularity after being published in "Scientific American" in 1970. This class of algorithms can exhibit extremely complex behavior on the basis of comparatively simple initial rules. Cellular automata are less suited for applications of style imitation, but allow for captivating results in the field of genuine composition.

Neural networks were originally created for tasks of image recognition. Their conceptual relation to biological information processing is one reason for their great popularity. In algorithmic composition, neural networks may generate outputs, whose sequences of note values need not necessarily occur in the underlying corpus. Disadvantages are found in the treatment of musical material that is context-sensitive over long passages, which can be better handled in generative grammars. For applications of algorithmic composition, neural networks are often used in the context of hybrid systems; interesting approaches can be found for neural networks in regard to suitable representations for musical information.

Artificial intelligence encompasses a variety of procedures for different tasks. Programs such as "ELIZA" or the well-known test by Alan Turing raise the question of a definition of the term "intelligence." Among other things, this chapter attempts to classify a number of procedures within artificial intelligence with regard to their relevance for algorithmic composition. Apart from single approaches, such as rule-based systems, alternative forms of logical reasoning, or variants of machine learning, it is above all the different forms of knowledge representation that are of the utmost importance for systems of algorithmic composition.

By means of a **final synopsis**, the last chapter is devoted to the basic motivations for algorithmic composition. It compares the properties of the treated paradigms, outlines principal strategies of encoding, representation and mapping, and finally deals with the limits of algorithmic composition and the attempts to establish systems of transpersonal validity.

The selection of the works covered was motivated by their exemplary character, an interesting approach, or their historic importance; naturally, they were also chosen according to the author's personal preferences. The richness of material available made it difficult to choose – thus, essential contributions may well have been overlooked and only particular aspects of the treated works could be described; also, any errors in content are solely the responsibility of the author.

References

1. Cambridge Advanced Learner's Dictionary 2006. http://dictionary.cambridge.org/define.asp?key=2032&dict=CALD. Cited 17 Jan 2006
2. Encyclopedia Britannica Online 2006. http://cache.britannica.com/eb/article-9005707. Cited 17 Jan 2006
3. Mazzola G (2003) The topos of music: Geometric logic of concepts, theory, and performance. Birkhäuser, Basel. ISBN 3764357312
4. Oxford Advanced Learner's Dictionary 2006. http://www.oup.com/oald-bin/web_getald7index1a.pl. Cited 17 Jan 2006

Chapter 2
Historical Development of Algorithmic Procedures

2.1 Interdependencies

A number of factors determined by, among others, sociology, social policy and philosophy, have created the basic conditions for the development of algorithmic thought processes and the instruments for their implementation. Aesthetics of different fields of art and their adoption in the course of history have also contributed essentially to their development. Accordingly, the intellectual attitude of an era becomes manifest in different disciplines, resulting in a number of interdependencies between different sciences. In the following, these relations are illustrated in the context of the period comprising the Baroque and the beginning of the Age of Enlightenment.

The establishment of natural science as an independent discipline has its origins in the Renaissance when this field freed itself from theology, and then made further advances in the Baroque period. Gutenberg's printing press with movable type, appearing around 1440, and the first runs of technical textbooks around 1550 were necessary preconditions for the wide transfer of knowledge. Innovations in seafaring led to drastic developments in precision mechanics, enabling among other things, the construction of the calculating machines designed by Schickard, Pascal and Leibniz.

In mathematics, this era marked the introduction of the analytical geometry of Descartes, the infinitesimal calculus of Newton and Leibniz, and Fermat's number theory. During this same period, the first textbook of probability calculus, the "Ars conjectandi," was written by Jakob Bernoulli. With Euler's "Introductio in analysin infinitorum," a ground-breaking work of analysis was created. These achievements are mentioned here only as examples; most of these mathematicians performed astonishingly in several fields, some of them independently achieving similar results, bringing up the question of first authorship, with perhaps the best known priority dispute which arose between Leibniz and Newton in the field of infinitesimal calculus. Logarithms too, for example, were independently discovered by the Scot John Napier and the Swiss Jost Bürgi at the beginning of this era. Naturally, these mathematicians also built upon the ideas of others. Single approaches of infinitesimal calculus, for example, can already be found in Archimedes' works. Number theory, too, was not first expounded by Fermat: It already had an important proponent in the ancient world in the figure of Diophantus of Alexandria.

The application of mathematical discoveries led to important developments in other branches of research: Galilei founded modern physics by systematically applying the experimental method. Similarly, Kepler's laws of planetary motion are a prime example of applied mathematics and form the basis for Newton's laws of gravitation.

As a result of these rapid developments, the scientist established his own independent statue. However, at the same time the increased specialization of the researcher became a more common trend and marked the end of the universal scholar as exemplified by Da Vinci as the personification of the Uomo Universalis. The dissolution of alchemy and chemistry by Lavoisier can also be seen in this era's area of conflict: alongside the scientific revolutions of the early Enlightenment, witch trials peaked and torture remained the primary instrument of establishing truth in judicial procedures.

Also comprising the Age of Absolutism, this era was characterized by the conspicuous display of power at European courts. The educational ideal of the Renaissance being made up of an agglomeration of educated scholars changed to an interest in technical apparatuses as part of courtly representation. Trick fountains like those constructed in Versailles, natural scientific cabinets where experiments were carried out for the entertainment of the guests of the court, collections of technical curiosa, theatre machines as well as a wide variety of different automata: all these became increasingly popular at courts. The entertainment value was the most important aspect – the more amusing and realistic the machine was, the more attractive it was for the ruler. The spectrum of devices appearing at this time ranges from pieces of furniture equipped with diverse technical extras to androids and artificial animals. Among the outstanding inventors were Jacques de Vaucanson, Pierre and Henri-Louis Jacquet-Droz and Wolfgang von Kempelen.

The construction of the automata was carried out by different trades. Watchmakers were responsible for fine mechanical motion sequences, millwrights were consulted for larger devices due to their knowledge of wheelworks, power transmission and leverages. Carpenters designed the body of the automaton and painters were responsible for its realistic appearance. By constructing automata for the courts, several trades were permitted to cooperate for the first time, since the ruler could abolish the mandatory guild membership.

The automaton as a technical innovation did not represent a scientific improvement in this context, but nevertheless required developments in the field of fine mechanics that could later on be applied in the realization of significant technical advancements. As an example, the invention of the pin cylinder may be mentioned here. It was first used for striking idiophones (such as small metal bars or bells), later being implemented in a modified form in Jacquard's loom and finally in Hollerith's punched card computer as a scanning apparatus. The automaton was also seen as a model of the biological body. In the dualistic view of Descartes, nature itself is nothing other than a complex mechanical apparatus. This view is also reflected in the materialistic-mechanistic anthropology of Thomas Hobbes. Proponents of the theory of rationalism arising at that time, including Leibniz, Wolff and Kant, base their assumptions on postulates such as metaphysics (the world is logical and constitutionally ordered) or epistemology (the world should be fathomed only by rationality, independent from human experience). These theses favor the presumption of the principal feasibility of all things. This view is opposed by empiricism with its proponents Locke, Hume and Berkeley, preferring immediate human experience as a means of knowledge.

The stages of development treated in the following, leading finally to the computer in today's sense, are naturally also affected by a number of determining factors and interdependencies that, however, cannot be covered here in detail.

2.2 Development of Symbol, Writing System and Numeral System

In order to be able to apply algorithms, the symbol must be introduced as a sign whose meaning may be determined freely, language must be put into writing and a number system must be designed.

The development of writing is generally known to be a three to four stage process. Based on a *logogram*, in a next step, a syllabary is created by means of phonetization that in a further differentiation finally develops to be an alphabetic writing system. Logograms can be *pictograms* – pictures resembling what they signify, *ideograms* – pictures representing ideas and serving as semantic carriers and *phonograms* – syllable signs. The phonogram may, but does not have to be, the starting point for an alphabetic writing system. For example, the chinese writing system consists of logograms that can also be found in Japan as *Kanji* along with the common writing systems. In general, the transition from a syllabary to an alphabetic writing system only occurs in languages that, due to their structure, can be further subdivided into their basic components.

The first fully developed writing system can be found around 3000 BC in Mesopotamia and Egypt. These writing systems are based upon originally pictographic representations that are later developed to reach a higher degree of abstraction. So, the long-winded notation of hieroglyphs changes to a standardized simplification, the *hieratic* system, that after further abstraction is finally replaced by the *demotic* system.

For the history of algorithmic thinking, an essential abstraction process has already taken place with the beginnings of these writing systems. An object may be represented by a sign and so it is possible to carry out abstract operations with objects, without using the objects per se.

It is also important to assign to the objects a number for their occurrence – meaning nothing more than the ability to count as well as creating a number system being capable of clearly representing the chosen values.

An interesting example of a division that has not been carried out between object and quantities is the counting method of the *Tsimshian*, an ethnic group in British Columbia, described by Tobias Dantzig [3, p. 6]. In this culture, there exist seven different term fields for numbers, depending on the objects associated with them. The Tsimshian distinguish between flat objects and animals, round objects and time, humans, long objects and trees, canoes, length specifications as well as objects that are not specified in detail.

This makes it clear that an essential precondition for the development of number systems – along with the ability to count – is, above all, the possibility to abstract the quantity of the related objects. In other words, it must be understood that four stones and four directions possess the same attribute, namely the number four as a mutual classification criterion. The example given above should only serve as a description of a possible process – an object-related number designation may have different causes and not only mean lacking abstraction ability, as it may possibly be assumed by an anthropologist from the Western world.

The oldest findings suggesting a conscious process of counting are given by bones with indentations. An examination under a microscope of the *Ishango bone* from Zaire allows interesting assumptions regarding its intended use: "This small bone, about 8000 years old, has a little piece of quartz stuck in a groove at one end and has three columns of notches cut into the sides of the bone itself. Microscopic examination of the bone indicates that the notches were made by some 39 different tools, thus it likely represents a record of events rather than a random decoration scratched into the bone. There is some limited evidence that suggests the markings may represent a record of some activity based upon a lunar calendar. If that is the case, then the notations record a series of events over a time span of almost six months. If the Ishango bone is really a numerical record rather than just a decoration, then traces of this type of recording system can be found from as early as the Upper Paleolithic cultures of 30,000 BC" [27, p. 39–40]. Similar findings that are much older are for example the *Lemombo bone* from the area between South Africa and Swaziland from around 35,000 BC as well as a finding from Vestonice in the Czech Republic, dated 30,000 BC.

Later, tally sticks are widespread and can be found in several cultures. Another interesting number notation can be seen in the *Quipu*, an arrangement of knotted strings of the Inca culture. With this system, also complex pieces of information may be encoded. Figure 2.1 shows the page numbers of the single chapters of the first edition of Donald Knuth's "The Art of Computer Programming," divided into volumes, in the form of a Quipu. The distance to the axe shows here the decimal position of the respective value given by one or multiple knots. In this case, the following division results: Volume I (454 pages) consists of chapter 1 (225 pages) and chapter 2 (229 pages); volume II comprises chapter 3 (155 pages) and chapter 4 (283 pages); volume III (541 pages) finally consists of chapter 5 (379 pages) and chapter 6 (162 pages). The last knotted string on the right side of the Quipu indicates the total amount of 1433 pages.

Provided that numbers are realized as an abstract value, further representation is made within a number system enabling a clearly arranged representation of larger

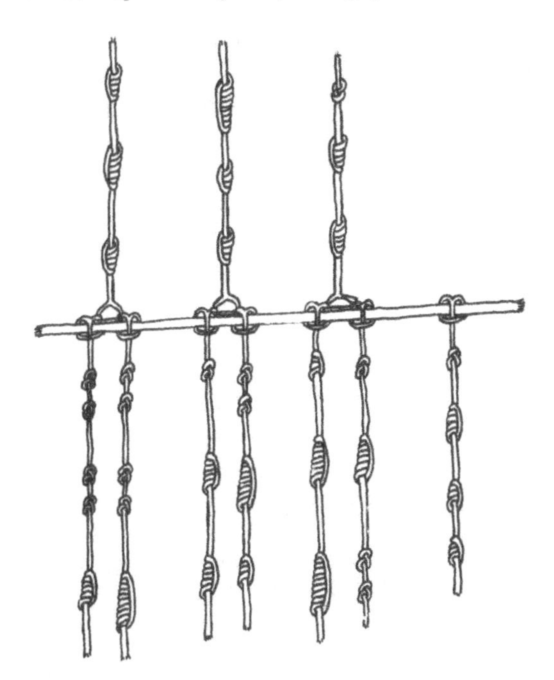

Fig. 2.1 Quipu of Knuth's "The Art of Computer Programming" [27, p. 40]. © 1986. Reprinted by permission of Pearson Education, Inc., Upper Saddle River, NJ.

values. Symbols for number units are generated whose main counting unit represents the basis of the number system. Systems on the basis 5, 10 or 20 can be found quite often, because of the anatomical correspondent given by the number of fingers and toes. Today, the decimal system has widely established itself. Still in today's use of language, however, evidence of older ways of counting can be detected. Traces of the supposedly Basque *vigesimal system*, based on units of twenty, can be found in French language in number designations such as "quatre-vingt" (four-twenty) for eighty or "cent-vingt" (hundred-twenty) for one hundred and twenty. Up to the 13th century, the denotations "six-vingt" (six-twenty) and "sept-vingt" (seven-twenty), amongst others, existed. In the Danish language, too, counting in units of twenty is applied, such as for example "tresindstyve" (three-twenty) or "firsindstyve" (four-twenty). Similar examples can also be found in English, Gaelic and other languages.

When comparing two of the first number systems, two basic systematizations may be recognized: Number symbols for different values are either cumulatively merged into units or symbols are used whose value is determined by their position. These systems are referred to as *additive* or *positional number systems*.

First sources of closed number systems can be detected on clay tables of the Sumerian empire around 3000 BC This *sexagesimal system* with sixty as a base was adopted by the succeeding Accadians and finally Babylonians. The basic number symbols consist of tens and ones that are merged in symbols up to the value 59. The position of the symbol in the symbol string gives information about the respective potency. Between the units of sixty, a little space is often left. Around 300 BC a placeholder is added for clarification in the form of two angular blocks whose function can be compared to a zero (see below), without, however, being an independent number which could be used for arithmetic operations. Similar functions are also performed by different symbols of the twenty-base vigesimal number system of the Classic era of the Maya from AD 300 to 900.

Fig. 2.2 Babylonian number 693, composed of 11 * 60 + 33 (arrows represent ones and hooks tens).

The Egyptian number system dates back to the time of the old empire (around 2700–2200 BC). The basic principle stays unchanged through all succeeding eras, although the hieratic (from around 2500 BC on) and the demotic writing system (from around 700 BC on) deriving from it enable an easier representation of numbers by merging multiply used symbols to single symbols. The hieroglyphic writing system is a decimal additive number system. The representation indeed starts depending on the notation (from right to left, from left to right or from top to bottom) with the smallest values; however, the potency of the base is only determined by the form of the hieroglyph and not by its position. In later times, a product representation is also used that, for example, represents the hieroglyph God (the sitting figure) once over five lines in order to express the number 5,000,000. The four basic

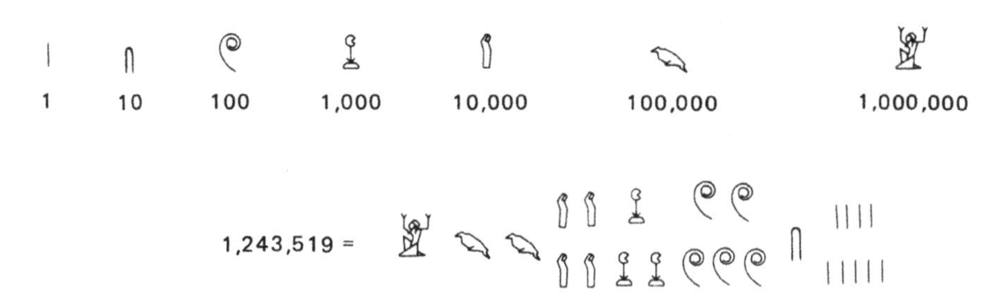

Fig. 2.3 Additive Egyptian number notation [27, p. 10]. © 1986. Reprinted by permission of Pearson Education, Inc., Upper Saddle River, NJ.

arithmetic operations are solved in Egyptian mathematics based on addition.

Well-known compendia comprising arithmetic problems were generated with the "Rhind mathematical papyrus" and the "Moscow mathematical papyrus" (from around 1700 BC on). The Greek notation developed in two differing systems, making use of either special symbols (from around 1100 BC on) or alphabetic characters (from around 500 BC on) to represent numbers. The Roman number notation is partly used even today; the Indo-Arabic number system became established in Europe, after some resistance, only from the 13th century.

2.3 Much Ado About Nothing – The Development of the Zero

Along with the introduction of today's number notation, there was a particular number gaining in importance that helps to profit from the advantages of positional systems in the first place. Not considering the placeholders of the Babylonians and Mayas mentioned above, the number zero as an independent number was developed by the Indians and first handed down through an inscription from Gwalior about 400 kilometers south of Delhi.[1] The inscription gives annualized the date AD 876 and describes the size of a garden. In the measure of length, a superior circle undoubtedly stands as a symbol for the zero. Most likely, the zero was used much earlier in Indian mathematics. With the campaigns of Alexander the Great, amongst others the Babylonian number system found its way to India that at this time used a number system similar to the Greek system. Around the year 500, the number system changed under the influence of Indian mathematicians. The Babylonian system was transformed into a decimal system and since 662 at the latest it was put down by a Syrian bishop that the Indians counted with new numbers. It is unknown whether the zero was already used at this point; however, it is most likely that the zero was used already before the inscription of Gwalior. The achievement of the Indian mathematicians was giving the number zero in addition to its function as a placeholder an independent position amongst the numbers – the zero became a value that could be used for calculations. The reasons for the development of the zero as an independent number in India and not gaining ground in Europe for such a long time can be found in differing philosophical and religious concepts. The zero being also a symbol for nothingness meets parts of Indian philosophy, which understands emptiness as the origin and objective of every development. The Sanskrit designation for the zero is "sunya," a term that simplified may be translated as "emptiness." In reality, the simple translation of "sunya" as "emptiness" is technically incorrect, since this emptiness may very well have content.[2] The introduction of the zero as an expression of a philosophical concept is, however, also opposed by another view that originally conceives the zero as the beginning of a new numeral series and therefore representing something similar to "10."

The Indian number system was introduced to Islamic culture by al-Khwarizmi around 820; "sunya" became the Arabic "sifr," latinization turned it into "zephirus," a word which the English term "zero" and the French term "zero" derive from. "Sifr" was also adopted as "cifra," a term from which eventually also the English "cipher," the French "chiffre" and the German "Ziffer" can be deduced from. A

[1] The information given in this section refers mostly to two works treating exhaustively the development of the number zero: Robert Kaplan: "The Nothing That Is: A Natural History of Zero" [13]; Charles Seife: "Zero. The Biography of a Dangerous Idea" [23].

[2] Interesting remarks on the definition of this term can be found in [13, p. 69ff].

long time before it was accepted in Europe, the Benedictine monk Gerbert of Aurillac (945–1003) made himself familiar with the Indo-Arabic numeral notation in Spain around 968. Gerbert's efforts to introduce the concept of the zero in the Christian occident met strong resistance in clergy. The reasons for this refusal have their earliest roots in the Pythagorean conception of numbers (around 580–500 BC). According to the Pythagorean system of thought, there is equivalence between numbers and forms; therefore a cube with a side length of zero loses its shape, and a relation including numbers that contain a zero does not make sense anymore. For the cosmos of the Pythagorean School which only expresses itself in number proportions, the zero therefore poses a threat – an invasion of nothingness, of chaos into a perfectly designed system. Pythagoras' approach, however, is controversial: The atomists postulate an empty space between the smallest components of the world. Zenon's paradoxon of Achilles and the turtle can only be solved by a limiting value of zero. The philosophy of Aristotle (384–322 BC), who adopted the Pythagorean views into his system, however, remained formative for the occidental thinking for nearly 2000 years. The zero was also declined by Christian theology that, influenced by Aristotelian philosophy, also denied the existence of emptiness and the infinitely small.

Fig. 2.4 Pope Sylvester II. © akg-images.

Gerbert's first attempt to introduce the new number system in Europe was not successful. In 999, he was consecrated Pope Sylvester II, but due to this attempt laid himself open to attack from the church. In France in the period of the Enlightenment, his coffin was opened to examine whether it not also contained a devil. In the 13th century, Saint Thomas Aquinas announced that God is unable to create neither an infinitely small nor an educated horse. However, an assembly of scholars, gathering in the year 1277 and chaired by the bishop of Paris, criticized the Aristotelian principles opposing God's almightiness – God's almightiness prevailed over the Aristotelian view of the world. Consequently, God can create the infinitely small, a vacuum. Yet half a century earlier, there had been efforts in Italy to introduce the zero.

Leonardo da Pisa, also known as Fibonacci, published his "Liber Abaci" in 1202, presenting a numerical series in which the following number is generated by the sum of its two preceding numbers, such as for example: 1, 1, 2, 3, 5, 8, 13, 21, etc. This numerical series also contains an increasingly exact representation of the *golden ratio* through the relation of two consecutive numbers. Geometrically, the golden ratio refers to the division of a quantity such that the ratio of the larger part to the whole quantity is the same as the ratio of the smaller part to the larger part. The golden ratio can be found in nature in several proportions and is already for the Pythagoreans an expression of perfect harmony. So, it is no coincidence that they chose a symbol for their cult in which the lines are divided in the relationship of the golden ratio, the pentagram. Leonardo da Pisa's "Liber Abaci" did not only introduce this numerical series, later known as *Fibonacci series*, but also favored the Indo-Arabic number system for arithmetic calculations. Leonardo was the son of Pisa's trade representative for Arabic countries and in this environment the Italian traders soon recognized the advantages of the new number system. So, it was not only a change of opinion that finally led to the implementation of the zero and the new number system, but simply its clear advantage for trade calculations.

Writing and number systems are the basis for the abstract use of any kind of object. However, developments that finally lead to the computer as it is today also require a formal system in order to carry out operations with the created symbols; within a given context, these operations should then lead to utilizable results.

2.4 The Formalization of Thinking Processes

In the occidental tradition, the formalization of thinking processes began with the development of logic. The term itself derives from the Greek "logos" that can either have the meaning of "thinking," "word," "thought," "sense" or "rationality." Again, it is Aristotle who paved the way for occidental logic. Indeed, the art of argumentation by means of logic is also proved to have been used by Socrates and Plato, but Aristotle is the one who developed logic as a closed system and established it as a scientific discipline. His works on logic can be found in the "Organon" as well as partly in his "Metaphysics." The denominations "Organon" ("tool," "sense tool," "method") and "Metaphysics" ("after the physics") do not originate from Aristotle but from Andronicus of Rhodes, who published these two works in the first century CE.

The basis for logical reasoning is thinking, which is assumed to be in principle consistent. A statement (*conclusion*) may be made due to facts (*premises*) and their mutual concepts. Of course, the underlying facts must be consistent in the sense of an *axiom*, being a premise which is accepted as absolutely right and therefore does not require any further proof. According to Aristotle, an axiom is a sentence that is taken for granted as valid, but it may be the basis for a proof. However, an axiom according to Aristotle may, compared to today's understanding, also represent a practical basic principle.

The conclusion or the conservation of a true statement out of the given propositions is subject to the laws of *formal logic deduction* that subordinates the special to the general. A *syllogism*, understood as a three-part conclusion, looks according to Aristotelian logic as follows: The first premise or *major premise*, followed by the second premise or *minor premise* results in the conclusion. The first and second premises have one term in common with each other known as the *terminus medius* (M) which has a subject (S) defined in the minor premise as part of a quantity defined in the major premise. A predicate (P) assigned to this quantity can in the conclusion also be assigned to the subject over the terminus medius. A well-known example for a three-part syllogism is the following conclusion:

- All men (M) are mortal (P): major premise
- Socrates (S) is a man (M): minor premise
- Socrates (S) is mortal (P): conclusion

Aristotle assumed three preconditions to be the basic principles for the process of the establishment of the truth. *Law of identity*: The terms used have the same meaning. *Law of contradiction* (also *law of non-contradiction*): There is no statement that is both true and false at the same time. *Law of excluded middle*: Every statement is either true or false. Together with the *principle of sufficient reason*, stating that every true thought must be constituted by a thought whose truth is proved, these four statements form the four laws of classical logic. Although the principle of sufficient reason is attributed to Leibniz, a premise of similar content can be found in the works of Democritus (460–371 BC): Nothing happens without a cause, everything has a sufficient reason. In order to receive general statements, Aristotle applied two types of *induction*: *Imperfect induction* uses a number of particular statements to get to a general statement. *Induction by enumeration* that can also be attributed to deduction, starts with proving a characteristic for a certain number of elements of a group in order to then prove this characteristic to be true for all other elements of this group. Socrates used induction to infer knowledge as a general term by observing particular cases.

Aristotelian logic determined the development of occidental thinking; however, from around 600 BC, there was also a tradition of Buddhist logic in India. The following construct is an example of an inductive five-membered syllogism in Indian logic:

There is fire on the hill (thesis to be proved), because there is smoke on the hill (smoke concurs with fire); there is smoke in the kitchen, where there is fire (this shows the relationship by means of a verifiable example); there is smoke on the hill (the observed state in the kitchen is generalized and applied to the situation on the hill); there is fire on the hill (conclusion).

Logic experienced further developments by the philosophy of the *stoics* from around 300 BC. The stoa distinguished between object, theoretical image and linguistic sign. The statement became the smallest relevant part of a logical operation;

for the first time, particles, known as *junctures*, were introduced to connect logical statements. Boethius (480–525) wrote about syllogisms, translated Aristotle's works and strove for a synthesis of Aristotelian logic and stoic logic.

The works of Aristotle fell into oblivion and only became present again in the High Middle Ages in the course of a return to antique writers. The socio-political changes in this era, such as higher agricultural production, improvement of commerce and specialization of trades led to wealth and a higher life expectancy. The settlement of the Investiture Conflict resulted in the Secularization of the ruling power. At the end of the 11th century, the first universities emerged from the monastery and cathedral schools. Reading and writing no longer were privileges reserved to the clergy.

The era of Scholasticism began with Anselm of Canterbury (1034–1109), putting rationality as a means of achieving knowledge alongside faith. In the 13th century, Aristotelian philosophy became, above all with Albertus Magnus (around 1200–1280) and his student Thomas Aquinas (around 1225–1274), an inherent part of scholastic thinking. When Anselm of Canterbury (1033–1109) stated, "credo ut intelligam"("I believe that I may understand"), so this implied a clear evaluation of faith being superior to logical thinking. Scholasticism tried to legitimize faith by means of rationality – logical thinking was cultivated, but mostly was an instrument for supporting Christian principles of faith. Outstanding logicians of that time were among others William Shyreswood (1190–1249) and his student Petrus Hispanus (around 1205–1277), whose "Tractatus Logicae" (around 1230), later known as "Summulae Logicales," was a recognized text book of logic and had been published 200 times by the 18th century.

In the context of these developments, the introduction of logic only represents the basis for the first attempts to systematize knowledge and to manipulate it in different ways within a system. These abstraction processes are necessary preconditions for the development of systems in whose context different algorithms for the generation of musical structure may also be applied.

2.5 A Truth Machine from the 13th Century

A provocative approach in this direction was made by Raimundus Lullus (also Ramón Llull, 1232–1316) with his "Ars Magna" (AM). It is about nothing less than the mathematization of knowledge and accessing it by means of a machine for the production of logical statements. The revolutionary concept of this approach exists in the idea that true statements can be obtained by algorithmic combinations of accepted terms.

Each combination of the AM has an underlying alphabet of nine letters. The symbols from B to K are semantic carriers of expressions in different categories, such as divine attributes, categorical determinants, question words, subjects, virtues and vices, as represented in table 2.1. Three diagrams and an arrangement of movable concentric circles form the working aids of the AM. The first combination figure,

B	bonitas	differentia	utrum	deus	justitia
C	magnitudo	concordantia	quid	angelus	prudentia
D	aeternitas	contrarietas	de quo	caelum	fortitudo
E	potestas	principium	quare	homo	temperantia
F	sapientia	medium	quantum	imaginativa	fides
G	voluntas	finis	quale	sensitiva	spes
H	virtus	maioritas	quando	vegetativa	caritas
I	veritas	aequalitas	ubi	elementativa	patientia
K	gloria	monoritas	quomodo	instrument.	pietas

Table 2.1 Semantic fields of the "Ars Magna."

or diagram A, represents the divine attributes of the first category that can appear both as noun and adjective. The lines in the centre describe possible relations. If, for the AM, the attribute also represents something universal, the meaning changes through the combination. So, in the combination of two attributes, one automatically becomes subaltern, such as in "truth is wise," etc. Because God is all-embracing in the theological context, his attributes are also universal, therefore forming the basis for everything that exists.

Fig. 2.5 Diagram A of the "Ars Magna." © bpk / Staatsbibliothek zu Berlin.

The next combination figure, or diagram T (figure 2.6), also refers to the alphabet of the AM; however, the letters represent the categories succeeding the divine attributes, such as categorical determinants, questions, etc. Further semantic fields in diagram T are three triangles lying on top of each other, representing possible relations and manifestations of terms. The first triangle includes "difference," "concordance" and "opposition." The second one stands for "beginning," "middle" and "end," the third triangle symbolizes "the larger," "the smaller" as well as "the same."

Every triangle edge – the relations of the manifestations – again divides into several possible cases; for example "the end" (finis – assigned to G as categorical determination) may happen through "termination" (terminationis), "privation" (privationis) or "perfection" (perfectionis).

Fig. 2.6 Diagram T of the "Ars Magna." © bpk / Staatsbibliothek zu Berlin.

According to Lullus, the alphabet of the AM interpreted through the different categories symbolizes the essence of every possible question as well as every basic statement of scholastic thinking. Alongside the diagrams, the AM is also made up of an exhaustive text material, in which uniform definitions for basic terms ("nature," "unit," etc.) are determined, interpretation rules (which attributes may be assigned to which subjects, etc.) are described and examples of interpretation are offered. Using the AM is a process of interpretation that happens in the context of three letters of the alphabet. In an arrangement of movably supported circular discs (figure 2.7), a letter triple may be placed which examines a particular issue with regard to different aspects.

These three letters are interpreted in the context of diagram A (divine attributes and predicates), of diagram T (the succeeding semantic fields) or in a combination of both these diagrams. These possibilities are shown in the contemplation table (figure 2.8), in which an interposed T indicates that the following letters are to be interpreted in the context of diagram T.

However, a necessary precondition for the use of the AM is knowledge of the interpretation rules and definitions of the text part. So, in the case of the interrogative particle "if," for example, a negative and a positive answer should be formulated that are verified by means of further operations. If a term is combined with the interrogative pronoun "what," questions such as the following result: What is the term? What are essential components or manifestation of the term?

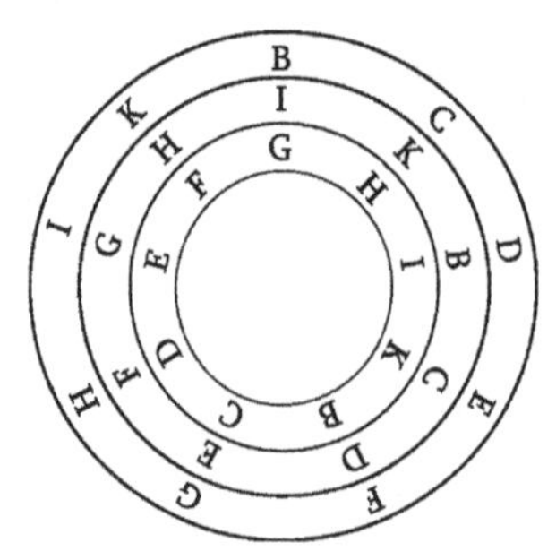

Fig. 2.7 Combination unit of the "Ars Magna."

Fig. 2.8 Contemplation table of the "Ars Magna." © bpk / Staatsbibliothek zu Berlin.

In the combination of semantic fields, it should be noted that, according to the text part of the AM, only allowed assignments are made, so that for example a vice cannot achieve the rank of a divine attribute, etc. Naturally, a combination of letters does not produce a particular question or statement, but opens up a whole field of discourse that is limited by the user in regard to a particular aspect. Namely these several possibilities of interpretation, that, however, are limited to a clear field of discourse by term definitions and interpretation rules, allow the user to consider his or her questions within a complex network of connected statements. In Lullus' own words: "The subject of this art is to answer all questions, provided that what man cannot know, can be formulated in the term [...] We have employed an alphabet in this art so that it can be used to make figures as well as to mix principles and rules for the purpose of investigating the truth. Through a letter that may carry several meanings, the intellect more generally comprehends manifold meanings and also knows them [...]." [1, p. 298].

Of course, the way of finding the truth in the AM can be disputed. The main problem is in the almost exclusive use of *circular reasoning*, meaning that what you are supposed to be proving is assumed to be true and the conclusion of the argument is implicitly or explicitly assumed in one of the premises respectively (also: *circulus vitiosus*), and consequently also in the breach of the principle of sufficient reason. In addition, the ambiguous contexts make clear statements nearly impossible. So in this system, truth can only be deduced in the context of the Christian dogmatism of that time, if at all. The underlying theological principles, however, are axioms according to Lullus and therefore an irrevocable basis of an objective establishment of the truth. His main objective is the conversion of the Islamic world by means of rationality, in the shape of the permutation discs of the "AM." At over eighty years of age, Lullus undertook his last unsuccessful mission voyage (1314–1316) and was stoned to death by an enraged crowd in Bugia, Algeria (figure 2.9).

2.6 Early Approaches to Algorithmic Composition

Also in the context of algorithmic composition, by around AD 1000 a first approach to automatic musical structure genesis had been developed. Guido of Arezzo (around 991–1031) contributed considerably to the development of notation, developed solmization and was an important music theorist of the medieval era. One of his most significant works, "Micrologus," is the earliest compendium of monodic and polyphonic singing practice. In chapter 15 and 17, he outlines a system for the automatic generation of melodies out of text material [20, p. 805]. Letters, syllables and components of a verse are mapped on tone pitches and melodic phrases (neumes), whereas groups of neumes are separated by caesurae. On the level of groups of neumes, the caesurae correspond to breathing pauses and can also be found in smaller groups in the form of pauses or held notes. The vowels in the text can be mapped on different tone pitches, as shown in figure 2.10. The concrete design of the melody is subject to musical limitations that are treated by Arezzo in his theory on "Motus."

In the time of Raimundus Lullus, a new musical genre was also becoming increasingly complex. The motet became the dominant form of polyphonic vocal music in occidental music history. At the beginning of this development was Leonin's (around 1136–1190) "Magnus liber organi de gradali et antiphonario," a compilation of chorales out of which through the further development of musical clauses complex polyphonic song forms eventually came into being. Beginning in the early 12th up to the early 13th century with Pérotin and Petrus de Cruce, the motet advanced through composers such as Philippe de Vitry (around 1291–1361), Guillaume de

Fig. 2.9 The stoning of Raimundus Lullus. Thomas Le Myésier: Breviculum ex artibus Raimundi Lulli electum; manuscript of the Badische Landesbibliothek Karlsruhe, Klosterbibliothek Sankt Peter (St. Peter perg. 92, fol. 10 recto). With kind permission of the Badische Landesbibliothek.

	A	B	C	D	E	F	G
a	e	i	o	u	a	e	i
o	u	a	e	i	o	u	a

a	b	c	d	e	f	g	a
e	u	a	e	i	o	u	a
u	i	o	u	a	e	i	o

Fig. 2.10 Mapping of vowels on tone pitches by Guido of Arezzo.

Machaut (around 1300–1377), Guillaume Dufay (around 1400–1474) and Josquin Desprez (around 1440–1521) and reached its last high point with Giovanni Pierluigi Palestrina (1525–1594) and Orlando di Lassus (1532–1594). From 1600 on, the motet lost its significance as a central musical genre, although it still appeared in the compositions of Bach, Schütz and other composers until the middle of the 18th century. The principle of *isorhythm*, invented by Philippe de Vitry and reaching its peak with Guillaume de Machaut consists of multiple repeating melodic (*color*) and rhythmic (*talea*) models that also interfere with each other and can occur in different proportions.

In order to meet the requirements of an increasingly complex polyphony, also a new concept of notation developed that, in contrast to chant and modal notation used until the early 13th century, also allowed the differentiation of rhythmic structures. Although in chant notation, developing from neumes, notes are arranged on single text syllables, the concrete rhythmic form cannot be notated with this concept. The modal notation, however, distinguishes between some triple-timed rhythms in different modes, although complex temporal structuring is still impossible with this notation form. Around 1280, the music theorist Franco of Cologne created his "Ars Cantus Mensurabilis," introducing a system of mensurated singing. The decisive innovation of mensural notation lies in the ability of the system to indicate the temporal duration of a note by its shape. In the mensural notation of the "Ars Antiqua" in Franco of Cologne's system and in the system of the "Ars Nova," influenced from 1322 on by Philippe de Vitry, the note values are parted in two or three. The values and the modes resulting from the divisions as well as accepted combination possibilities enable a complex rhythmic repertoire. The form of notation as known today came into being around the 16th century and stems from further developments and roundings of the mensural note values.

Both mensural notation and the complex musical procedures of the motet illustrate the essential abstraction achievements of this era, ones that are also of great importance to the development of algorithmic composition. Mensural notation enables the representation of several musical parameters with a symbol and constitutes an event space for possible rhythmic constellations with constraints of the permitted combination possibilities. The taleas and colors of the motet, however, show the structure generating application of musical parameter series – a procedure which came back into use in serialism of the 20th century.

If the abovementioned principles represent the beginnings of algorithmic manipulation of musical material, then the "Ars Magna" of Raimundus Lullus effectively realizes the concept of a computer (music) system. The analogies to hardware and software, data memory, program, etc. are evident in the components, definitions and rules of the "Ars Magna." Lullus creates a system that due to its underlying structure (hardware, corresponding to the diagrams), a knowledge base (data, corresponding to the definitions) as well as application instructions (software, corresponding to the interpretation rules) independently generates statements. Because of the given combination possibilities, some degree of chance is involved that, however, due to the interpretation rules, provides coherent statements in the given context whose exact interpretation is left to the user.

The circular statements inherent in the "Ars Magna" can actually also be found in a system of algorithmic composition, because any compositional premises can hardly be compared to axioms. When compositional work is considered under this point of view, the circular statement is as inherent in it as it is in the "Ars Magna." Of course, this objection can be met with a formally closed composition system that algorithmically manages every structural decision. The attempt to generate musically incontrovertible structure by applying proven or unproven sentences of any designed system on musical parameters is made time and again. Hauer's "Zwölftonspiel" or Schillinger's[3] composition system that is based on "mathematical legalities," are only two examples of transpersonalizing compositions by referring to extra-musical systems and, by doing so, making them unassailable. The formal closedness, or "truth," of any principle taken as a basis, becomes inevitably lost through the musical transfer. What is left is the concretely produced structure, the musical information.

The musical quality of a structure produced in this way (as well as in all other ways) has to be left aside, because even value judgments exerted by musicological discourse are inevitably subject to personal preferences or trends. The "Ars Magna" enables the user to make a number of interpretations within the clearly defined rule system. Here, too, a parallel may be drawn to a system of algorithmic composition that in most cases allows the generation of a whole class of compositions by producing a meta-structure.

Whereas, according to Lullus, combinatorics opens a space for self-reflections within a controlled system, a concept of the Jesuit and universal scholar Athanasius Kircher (1602–1680) allows the output of directly utilizable material in the musical context. Kircher worked in the fields of astronomy, mathematics, medicine, music, mineralogy and physics as well as in linguistics, where he tried to decipher the Egyptian hieroglyphs with combinatorial methods. Although his effort failed and the correct decoding was only completed by Jean François Champollion (1790–

[3] "The Mathematical Basis of Arts" [22] and "Schillinger System of Musical Composition" [21].

1832) with the Rosetta Stone, Kircher's attempts nonetheless represent a remarkable achievement in combinatorics and paved the way for applied cryptology.

Fig. 2.11 Atanasius Kircher. © bpk.

In his exhaustive musicologist work "Musurgia Universalis" from 1650, Kircher developed amongst others a system of algorithmic composition. This system consists of three categories of labeled wooden sticks (syntagmas) on which both numbers and rhythmic values are engraved. Kircher's system allows for the automatic generation of contrapuntal compositions in the style of the contrapunctus simplex and floridus. In an advanced form, style-typical material of particular musical genres can be produced.

In Kircher's "Arca Musarithmica," four-lined number columns can be combined with four-voice rhythmic patterns by means of the syntagmas. The number columns represent levels of different modes and are arranged in groups of 2 to 12 units. These units serve to correctly transfer text passages and represent one syllable each. Each class of tone pitch symbols of a particular size can be combined with a class of rhythmic patterns of the same size, finally producing four-voice movements in the style of the contrapunctus simplex. Because the number of voices differs in a movement of the contrapunctus flores, in this form of syntagma the voices are only combined with a selection of appropriate values.

In the so-called "Arca Musurgia" that, however, is only incompletely described in the "Musurgia Universalis," the tone pitches and rhythmic patterns are advanced in a way that they enable the composition of material within five classes of musical genre (church style, madrigal, motet, fugue, and monody).

The fact that in Kircher's system tone pitches are represented by numbers advances the class of possible compositions, since abstractions can be made from a concrete mode. Moreover, this principle shows an application of *pitch classes*, which have been a common representation system in the production and analysis of musical material since the 20th century.

Combinatorics is also used in another system of Kircher in order to solve tasks from various fields. In 1661, he created an apparatus for the twelve year old Carl Joseph of Habsburg that he named "mathematical organ." This device is similar to a card index system and treats arithmetic, geometry, the building of forts, ecclesiastic time calculations, sundials, astronomy, astrology, cryptographs as well as music according to the above mentioned principles. For each special field there is a collection of labeled discs whose use (the possible combinations for solving the task), are explained in an enclosed booklet. Not all possible procedures are created by Kircher; the arithmetic part, for example, uses *Napier's bones*, for geometrical tasks the construction of the *geometrical square* (a common instrument of surveyors made up of a square frame with two scales and a rod to measure distances) is applied. The importance of Kircher for the development of algorithmic thinking and finally the computer can be seen amongst other things in the fact that as an advancement of Lullus' concept, a mechanical arrangement generates concrete outputs that may directly be used for solving a problem. Because all solutions possible in the system are also coded by the combination possibilities of the system, Kircher also paves the way for a comprehensive representation of knowledge in a chosen field of discourse.

2.7 A Utopia of an All-Embracing Representation of Knowledge

A decisive approach in this direction was made by Gottfried Leibniz (1646–1716). Besides other essential contributions to the development of science, he also pursued a method for formulating all knowledge in a kind of "universal language" in order to enable the solution of any scientific problem within a comprehensive calculus. In a letter from 1679 asking for support for his plans, Leibniz wrote to his employer Duke Johann Friedrich of Braunschweig: "[My] invention uses reason in its entirety and is, in addition, a judge of controversies, an interpreter of notions, a balance of probabilities, a compass which will guide us over the ocean of experiences, an inventory of things, a table of thoughts, a microscope for scrutinizing present things, a telescope for predicting distant things, a general calculus, an innocent magic, a non-chimerical cabal, a script which all will read in their own languages; and even a language that one will be able to learn in a few weeks and which will soon be accepted amidst the world." [5].

For Leibniz, who often also signed himself "pacidus" (from Latin, meaning "conciliator"), the concept of a universal language was also a vision of cross-national communication in Europe after the Thirty-Year War. His program for the formalization of the sciences was to be achieved through a number of preconditions: First, by compiling a complete encyclopedia of all terms that are necessary to formalize the sciences; second, a formal system was to be developed, namely the "lingua universalis," within which all scientific terms could be coded; third, by creating a

Fig. 2.12 Gottfried Leibniz [4, p. 21].

calculus, the "calculus ratiocinator" that should enable the connection of all terms of the universal language by means of logical operations. Assuming the universal encyclopedia to be feasible, Leibniz searched appropriate symbols for the representation of all scientific terms in his universal calculus. One postulate was that by combining such symbols, a logical relationship occurred. As a set of symbols, he chose the group of natural numbers. All terms whose subject features one particular characteristic, must be represented by natural numbers that are whole-number factors of a number representing the *predicate* or, according to today's understanding, the *class*. Therefore, according to Leibniz' *characteristic numbers*, the following *heredity concept* (meaning here that every object belonging to a class takes on all characteristics of its superior classes) would result from the example of the object 'Englishman'[8, p. 10]:

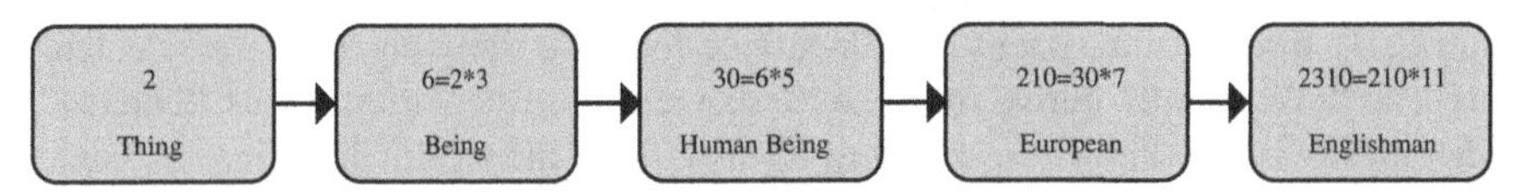

Fig. 2.13 Characteristic numbers by means of an object hierarchy.

Another logical connection could be of the following form: Animal gets the number "2," the predicate "rational" number "3," consequently for a human as an intelligent being the result is number "6" that can both be divided by three and two. The characteristic numbers that cannot be divided any further must naturally be represented by prime numbers in order to enable explicit partition relations. Although it is a groundbreaking approach, the concept of characteristic numbers also exhibits a few problems regarding the application of some term attributions as well as the representation of combinations of Aristotelian logic. Leibniz himself has carried out some advancements of his characteristic numbers.[4]

The concept of an ideal language is also considered by Descartes; for him, too, numbers represented the best possible symbols of a universal vocabulary. Leibniz, however, elaborated a concrete outline of a system, regardless of the impossibility of its realization. Even though through modern logic some aspects of Leibniz' program are realized, the works of Gödel and others (see below) show the limits of decision making in logical calculus. In Leibniz, numbers become an instrument for organizing the world – even music results from a metaphysical act of calculating: "Musica est exercitium arithmeticae occultum, nescientis se numerare animi" ("Music is a hidden arithmetic exercise of the soul, which does not know that it is counting."). According to another well-known quotation of Leibniz "Dum calculat Deus, fit mundus" ("When God is counting, a world comes into being."), counting becomes a religious creation process. The fact that an attention to formal concepts may also come with a deep sense of spirituality is also shown by a creation of Leibniz' prominent contemporary J.S. Bach, who combines in The Art of Fugue or in his Goldberg variations complex structuring procedures with highest musical expression.

2.8 Calculating Machines

Leibniz' reflections concerning a comprehensive calculus also include the possibility of mechanized calculation. Due to the progress in fine mechanics, which saw continuous improvement because of the rapid development of automata in this era, it was possible for Leibniz to design a prototype of a functioning calculating machine. Indeed, its development had a long tradition, going back to the ancestor of all calculating machines, the abacus. The principle is quite simple: Numbers are represented according to their place value by balls or stones in different columns. By moving the units, addition, subtraction and also combined calculating operations can be carried out. Early forms of the abacus can already be found in Babylon. The terms "calculus" and "to calculate" derive from the Roman term for calculation stones, "calculi." The abacus is also known in differing forms around the world: In China as "Suan Pan," in Russia as "Stchoty" and in Japan as "Soroban." A form of abacus that was used in the Middle Ages, on which marks for the representation of number units were moved on different lines, is another version of this successful principle.

One of the first actual calculating machines goes probably back to an idea of Leonardo da Vinci (1452–1519). The device described in a draft has a number of cog wheels that are capable of moving other cog wheels by one position after ten turns. With this principle, it is possible to carry out additions in the decimal system; however, it is not proven if this apparatus should represent a calculating machine or a system for gear transmission.

[4] Further examinations of the concept of characteristic numbers as well as approaches to the mentioned problems can be found amongst others in [8].

A mechanization of calculating processes, though not by a conventional calculating machine, was already described in 1617 by the Scottish mathematician John Napier (1550–1617) in his publication "Rabdologiae." Napier, who, along with Jost

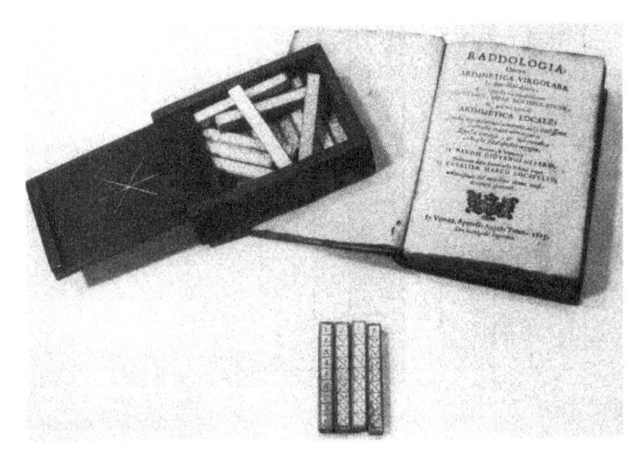

Fig. 2.14 Napier's rods and title page of a calculating instruction by Lord Napier, translated into Italian, Verona 1623 [4, p. 17].

Bürgi (1552–1632) pioneered in the field of logarithms with his "Mirifici logarithmorum canonis descriptio" appearing in 1614, designed a device for multiplication and division, based on the principle of the *Pythagorean abacus*. The Pythagorean abacus is a tool consisting of a table for reading off the times multiplication table. *Napier's abacus*, also known as Napier's bones, or Napier's rods, consists of rods with product rows printed on each side where the digits of the numbers are separated by diagonal lines. Figure 2.15 shows this principle with the help of the multiplication of 2357 * 7. In the seventh line of the number "2357" placed in the first row, are the products with factor "7." The result comes from the addition of the digits within the diagonals. For multi-digit multiplications, the partial products are added. In division, the quotient is obtained stepwise by indicating all products of the respective divisor.

Caspar Schott (1608–1666), Jesuit padre and professor of mathematics at Würzburg University, described in his "Organum Mathematicum" of 1668, containing a detailed description of the "mathematical organ" of his teacher Kircher, a mechanical arrangement of Napier's bones. In principle, this device consists of pivoted cylinders on which Napier's bones are placed. Each of the openings on the upper side of the housing makes a column visible. Due to its construction, Schott's "counting box," however, is more a representation of a procedure than an actual calculating machine.

Another predecessor of mechanical calculating machines is an invention of the English astronomer Edmund Gunter (1581–1626) in 1623. Gunter combined the concepts of Napier's logarithms and the calculating rods of the English mathemati-

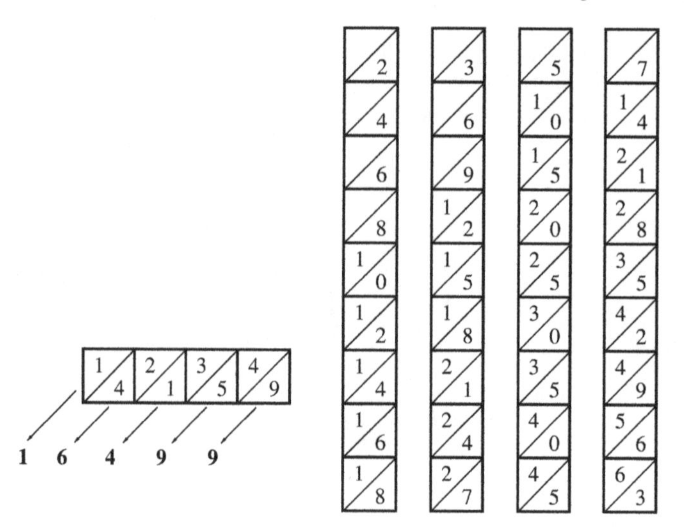

Fig. 2.15 Multiplication with Napier's rods.

cian William Oughtred (1574–1660) and developed the basic concept of the slide rule that was still used in the second half of the 20th century.

The first model of an actual calculating machine that also used the principle of Napier's bones was designed by Wilhelm Schickard (1592–1635), professor of astronomy at Tübingen University, around 1623. For addition and subtraction, the principle of the decadic wheel was used, where after a complete turn the wheel of higher value raises one notch. In the Thirty-Year War, an eventual prototype was lost; however, the function principle has come down to us through Schickard's correspondence with Kepler and allows for a fully functioning reconstruction that was built by Bruno Freiherr von Freytag-Löringhoff, professor of philosophy at Tübingen University. This reconstruction of Schickard's calculator became popular in 1973 through its picture on a stamp designed on the occasion of the 350th anniversary of this invention.

Between 1642 and 1645, Blaise Pascal designed a number of prototypes of his "Pascaline." Like Schickard's calculating machine, this construction, too, consists of rolls with an automatic ten carry mechanism.

For multiplication, in the machines of Schickard and Pascal, first the partial products have to be formed so that they can be added together by the mechanism. The Pascaline supports only addition of partial products; in Schickard's machine, these can be read off from Napier's rods.

Other calculating machines containing a mechanism for addition or using Napier's bones for multiplication were built by Samuel Morland (1673) and René Grillet (1678).

A significant step of development was made by Leibniz, who enabled multiplication through automatic multiple additions with his invention of the *step reckoner*. The step reckoner is a variable-toothed roll that can move another wheel by up to

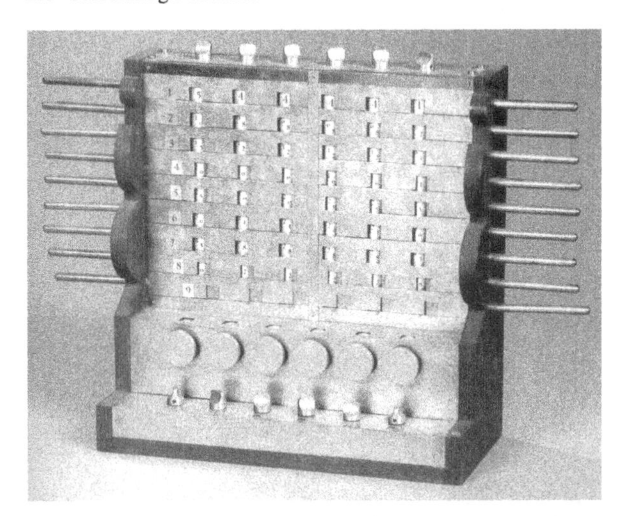

Fig. 2.16 Reconstruction of Schickard's calculating machine. Picture courtesy of the archives of the Technisches Museum Wien. Reproduced with kind permission of the Technisches Museum Wien.

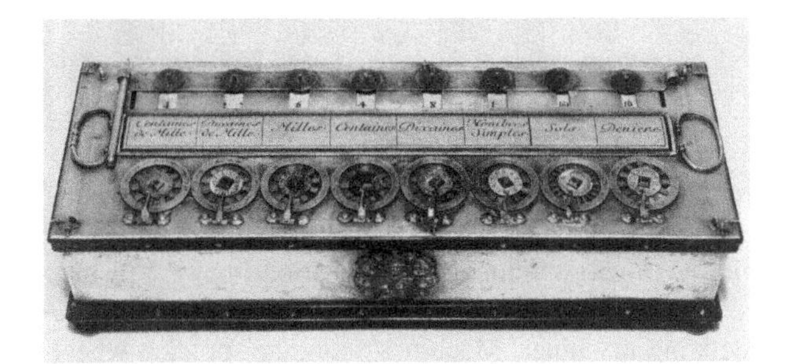

Fig. 2.17 Pascaline. With kind permission of IBM Corporate Archives, Somers, NY.

Fig. 2.18 Calculating machine by Grillet. Photo Herbert Matis. With kind permission of Herbert Matis.

nine notches, depending on its position. Leibniz used this concept in his first prototype, completed in 1673; the automatic ten carry mechanism remained at this time a nearly intractable mechanical challenge. Other well-known calculating machines based on the system of the step reckoner were built by the Protestant priest Philipp Matthäus Hahn as of 1770 and by Johann Helfrich Müller between 1783 and 1784.

Fig. 2.19 Leibniz' calculating machine on the principle of the step reckoner. View from below at the mechanism [4, p. 21].

Another significant construction based on Leibniz' ideas is the sprocket wheel which was invented by the Italian mathematician Giovanni Poleni around 1709. Consisting of a cog wheel with teeth whose number can be varied from the inside, it enables the same functions as the step reckoner and could be found up to the late 20th century in different calculating machines. Machines using the sprocket wheel were also built by the German watchmaker Antonius Braun around 1727 and the German mechanic Jakob Leupold around 1750.

Fig. 2.20 Calculating machine by Leupold [4, p. 24].

The first mass-produced calculating machine was the "Arithmomètre" of Xavier Thomas de Colmar (1785–1870) patented in 1820 and built in a run of over 18,000 pieces. Further developments such as the proportional lever, making the ten carry mechanism easier, or the direct multiplier enabling multiplication in only one step, led to continuous improvements of mechanical calculating machines and finally reached their fine mechanical zenith with the "Curta." This handy four operations

Fig. 2.21 Curt Herzstark's "Curta." Photo Herbert Matis. With kind permission of Herbert Matis.

machine (all basic arithmetic operations can be carried out) was developed by the Austrian producer of calculating machines Curt Herzstark from 1935 on and is the last mechanical competitor of the pocket calculators, appearing in the 1970s.

2.9 A New Numeral System for Automated Calculations

Leibniz' reflections on the optimization of mechanical calculations also led a lot further. In 1679, Leibniz' considered a new system for mechanical calculations: "This [binary] calculus could be implemented by a machine (without wheels), in the following manner, easily to be sure and without effort. A container shall be provided with holes in such a way that they can be opened and closed. They are to be open at those places that correspond to a 1 and remain closed at those that correspond to a 0. Through the opened gates small cubes or marbles are to fall into tracks, through the others nothing. It [the gate array] is to be shifted from column to column as required." [12, p. 46–47]. The machine itself was never built; the concept of the newly invented number system, the binary system, was only formulated 22 years later in 1701 by Leibniz in his work "Essay d'une nouvelle science des nombres." Leibniz' own evaluation of the new system does not seem exaggerated, regarding its relevance for later developments: "Despite its length, the binary system, in other words counting with 0 and 1, is scientifically the most fundamental system, and it permits new discoveries in [...] geometry, because when numbers are reduced to the simplest principles, like 0 and 1, a wonderful order appears everywhere." [7]. Leibniz, an admirer of Chinese culture and striving for an intellectual exchange, described his system in 1701 in a letter addressed to Jesuit padres Joachim Bouvet and Claudio Filipo Grimaldi, who worked as mathematicians for the Chinese emperor. In November of the same year, Bouvet answered Leibniz with reference to the binary structure of the "I Ching," [5] whose different *hexagrams* (in this context, a symbol consisting of six lines) are only represented by two symbols. Although all hexagrams in the "I Ching" are made of only two symbols, they cannot be used for representing information outside of the concrete context. Leibniz, however, designs his binary system or the *dyadic* as an independent numeral system capable of carrying out all kinds of calculations. Moreover, the lines in the "I Ching" can get an additional dimension through the so-called *change*. Depending on the hexagram, the *female* (broken) or *male* (unbroken) *lines* can also appear as *changing lines* that change to be their opposites in the symbol that is interpreted next. In this respect, the set of hexagrams of the "I Ching" consists strictly speaking of four symbols; therefore, also this system can hardly be considered an early model of a binary system.

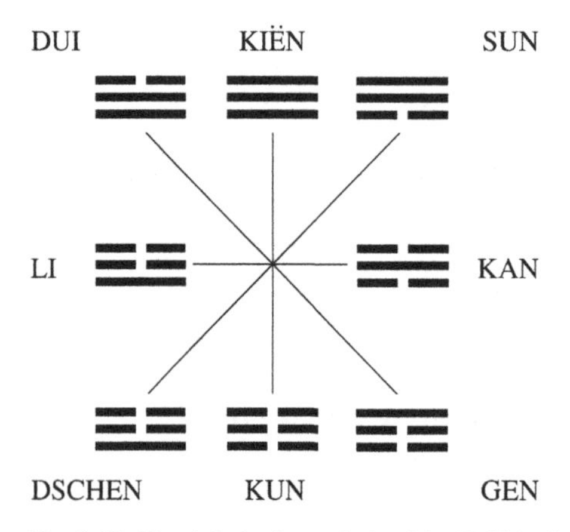

Fig. 2.22 The eight basic symbols of the "I Ching" for the construction of the 64 basic hexagrams.

2.10 Replacing the Mechanistic Determinism

Leibniz' reflections on a comprehensive calculus, however, also include another essential thought: "I have said more than once that we need a new kind of logic, concerned with degrees of probability. [...] Anyone wanting to deal with this ques-

[5] For detailed information, see Wilhelm's descriptive book [26].

tion would do well to pursue the investigation of games of chance. In general, I wish that some able mathematician were interested in producing a detailed study of all kinds of games, carefully reasoned and with full particulars. This would be of great value in improving the art of invention, since the human mind appears to better advantage in games than in the most serious pursuits." [16, p. 466]. The fact that Leibniz included probability calculus and game theory, developed in its modern form in the 20th century by John von Neumann and Oskar Morgenstern, in his reflections, emphasizes that his concept is effectively pointing the way to a comprehensive calculus. Amongst others, his reflections comprise the following objectives:

- The construction of a language-independent representation system as a knowledge basis through the concept of his universal language.[6] Within this system, the logical relations between the objects can be seen directly from their symbols; in the case of the universal language, this would mean the partition ratio of the numbers representing the objects.
- The attempt to mathematize logic in order to calculate truth values; this should consequently lead to achieving mechanical decision-making within the calculus.
- The draft of a numeral system optimized for automatic applications, namely the binary system.
- The integration of probability grading in the process of the logical establishment of the truth.
- Analysis of action strategies in systems with given rules by means of game theory.

The integration of probability theory in a calculus enables on the one hand reasoning in arguable situations, on the other hand a significant reduction of data. With Leibniz' own works in the field of infinitesimal calculation, whose basic concepts Newton also develops, the view of *mechanistic determinism* is also assured, propagating an unchangeable course of world affairs and therefore opposing the concept of probability. In this sense, for Pierre Simon de Laplace (1749–1827) knowing all starting states and rules of a system also includes the possibility of being able to exactly predict all future states.

This concept of classical physics became obsolete at the latest with the works of Werner Heisenberg (1901–1976), who states in the closing section of his work on the indeterminacy relation: "If we know exactly the present, we can predict the future. In view of the intimate connection between the statistical character of the quantum theory and the imprecision of all perception, it may be suggested that behind the statistical universe of perception there lies hidden a "real" world ruled by causality. Such speculation seems to us – and this we stress with emphasis – useless and meaningless. For physics has to confine itself to the formal description of the relations among perceptions." [11, p. 172].

[6] Cf. Umberto Eco's "In Search for the Perfect Language," which traces the history of a universal language [5].

2.11 Language and Music Generators – A Book of Books

Whereas Lullus related abstract qualities by combining circular discs, Georg Philipp Harsdörffer (1607–1658) generated (with a similar arrangement) concrete material in a linguistic context. The "Fünffache Denckring der Teutschen Sprache" ("Five-fold thought ring of the German language") is an arrangement of concentric circles by whose combination according to Harsdörffer the entire German language can be represented on one page (figure 2.23). The mechanical principle is identical with Lullus' arrangement, but here, the linguistic combination material is placed on five rings. Harsdörffer structured this material in so-called prefixes, initial letters and rhyme letters, middle letters, end letters and suffixes. Of course, not every generated unit is automatically part of the language. The user is responsible for the examination of the results and so first must know the correct lexicon to be generated in order to be able to successfully work with this machine. The generative power of Harsdörffer's thought ring is unlimited, and it is notably this fact that makes its outputs arbitrary, because necessarily the examination of the results is left to the user.

In the 18th century a game became popular where the outputs were also generated on the basis of combinatorial possibilities but which worked completely independently of the evaluation of the user. This was a new musical amusement which also made it possible for a layperson to produce compositions in different musical genres. Mostly, the user chose for every temporal unit a bar out of a particular table by rolling a dice, until a short piece of music of a musical genre had been produced. The principle of the so-called "musical dice game" is simple, but effective: For each temporal position (mostly bar positions), a number of musical constellations (mostly bars) must be available out of which one can be chosen deliberately without running the danger of producing musical clashes by doing so. For example for bar 1, a number of musical arrangements of the tonic are possible, for bar 2 arrangements of the subdominant are used and for every further bar again a number of options is available. The process continues until finally the required number of bars is reached. The number of necessary bar variations complies with the possible numbers on the dice; with two dice, the most common application, 11 variations must therefore be available. The application of the musical dice games is trivial; however, its design requires some talent in composing music, because for all musical possibilities attached one to another, not only the harmonic but also technical aspects of voice-leading have to be taken into consideration. The first dice game is Johann Philipp Kirnberger's "Der allezeit fertige Menuetten- und Polonaisencomponist" ("The ever-ready minuet and polonaise composer"), designed in 1757; up to the year 1812, at least 20 other creations of this type were built. Dice games of well-known composers are "Einfall einen doppelten Contrapunct in der Octave von sechs Tacten zu machen ohne die Regeln davon zu wissen" ("A method for making six bars of double counterpoint at the octave without knowing the rules") by Carl Philipp Emanuel Bach in 1758 or "Table pour composer des minuets et des Trios à l'infinie; avec deux dez à jouer" ("A table for composing minuets and trios to infinity, by playing with two dice"), created around 1780 by Maximilian Stadler. From

Fig. 2.23 To the bookbinder. This figure has to be cut out, separated into five circles and fixed on five discs of the same size, so that every circle can be moved separately. After that the fivefold ring can be put in again.
The "Fünffache Denckring der Teutschen Sprache." Harsdoerfer, G.P.: Delitiae mathematiceae et physicae ... Nürnberg, Dümlern, 1651. Research Library in Olomouc, sig. 619.407. Reproduced with kind permission by the Research Library in Olomouc.

1793 on, also musical dice games under the names of Haydn and Mozart appear; however, their having created these games has not been proven.[7]

Harsdörffer's thought ring defines a complete space of linguistically possible word creations. Procedures combining existing material or constructing new material with limited parameters can in literature also lead to the formation of new fields of artistic expression. *Anagrams* (generation of new words by rearranging letters), *palindromes* (text reading the same in either direction), *lipograms* (text avoiding certain letters) and many more are common procedures for the structural treatment of text material. Beginning with one of the first palindromes "Sator Arepo Tenet Opera Rotas," first documented by archaeological excavations in Pompeii, up to the different methods of rule-bound literature of Raymond Queneau, the tools applied have a wide range. Queneau was a founding member of "Oulipo" (short for French "Ouvroir de Littérature Potentielle") whose members aimed at expanding literature by means of formal principles. Two examples of Queneau are now demonstrated. In "cent milliards de pomes," for a sonnet comprising a total of fourteen lines, 1014 possible sonnets can result from ten different verse lines each. With "un conte à votre façon" from 1967, he created a branching system enabling multiple readings of a story through freely selectable segments, similar to the structure of a *hypertext system*. Such a system uses non-linear connections between texts and was first designed by Vannevar Bush in 1945.

A well-known example for a lipogrammatic novel is "La disparition" by Georges Perec, also member of "Oulipo." The text consisting of 85,000 words altogether avoids the use of the letter "e."

Combinatorics as a tool for producing a "book of books" also represents a popular object of speculation. Mallarmé's draft "Le livre" should – as a work that can be used in different combinations – represent a universal book, dissociated from its author, of all potentially possible pieces of literature. These approaches can also be found, taken more or less seriously, in the works of Jorge Luis Borges,[8] Kurd Lasswitz[9] and others. The already mentioned "I Ching" best meets this claim. Because with the changing lines each of the 64 basic symbols may pass over to another symbol (in resting lines also to themselves), 4096 (64^2) possible symbol combinations can result. Because every basic symbol also has texts for all possible lines in their changing states, every combination of symbols can also be read as a text. However, the "I Ching" is not universal in the sense of Mallarmé, because it naturally stays limited to its context.

[7] Haydn: "Gioco filarmonico o sia maniera facile per comporre un infinito numero de minuetti e trio anche senza sapere il contrapunto" ("The game of harmony, or an easy method for composing an infinite number of minuet-trios, without any knowledge of counterpoint"); Mozart: "Anleitung zum Componieren von Walzern so viele man will vermittelst zweier Würfel, ohne etwas von der Musik oder Composition zu verstehen" ("Instructions for the composition of as many waltzes as one desires with two dice, without understanding anything about music or composition") as well as "Anleitung zum Componieren von Polonaisen..." (Instructions for the composition of polonaises..."). For the history of the dice games in the 18th century, see the detailed article by Stephen A. Hedges: "Dice Music in the Eighteenth Century" [10].

[8] "The library of Babel" [2].

[9] "Die Universalbibliothek" [14].

Fig. 2.24 Stephan Mallarmé. © akg-images.

The principles of literature based on rules form a field of general procedures that also occur in a musical context. Structural parameters or conditions for the application of particular material can be found in numerous compositional methods, above all in the 20th century. Twelve tone technique, serialism and aleatorics are only some of the methods for musical structure genesis. The variable reading of Queneau's "un conte à votre façon" has its early parallel in Stockhausen's piano piece XI from 1956, in which a pianist constantly finds new paths through 19 fragments of the score. But the "I Ching," too, finds its way into composition. John Cage's "music of changes" from 1951 generates its pitches and durations out of the symbols of the "I Ching" that according to a traditional method are selected by threefold coin flipping.

2.12 From the Loom to the "Analytical Engine"

Leibniz' idea of applying the binary numeral system to an automaton was realized by an invention in the textile industry. In the year 1725, Basile Bouchon developed the first system for an automatic control of looms. Information about raising and lowering the warp threads is transmitted via punched holes in paper tapes. In 1728, Jean Baptiste Falcon replaced the fragile paper tapes with punched wooden boards. The boards, like the paper tapes, must be fed manually; an automatic control mechanism was only developed by the automata constructor Jacques de Vaucanson in 1745. In 1805, Joseph-Marie Jacquard built the first model ready for production; by 1812 already 10,000 models were in use in France.

The concept of scanning punched holes to transmit information was later used for the control of different automata. This basic principle of the punched card as a control and storage medium was only replaced by the magnetic tape in the 1960s.

Fig. 2.25 Loom after Jean Baptiste Falcon. © bpk.

On the basis of these concepts, Charles Babbage (1791–1871) was the first to tackle the construction of a universal calculating machine. Babbage was a mathematician and political economist. As an economist, he published amongst others "On the Economy of Machinery and Manufactures" in 1834 whose market-economical analyses and reflections on production and factory system triggered the reorganization of the English economic system on a scientific basis. Babbage was a member of the "Royal Society" and "Lucasian professor" at Cambridge University. His mathematical works also set the life insurance business on a scientific basis. Babbage's main objective was the mechanization of calculation processes such as those that are required for the generation of logarithm tables. Because difference calculation may be used to trace back complex expressions to continuous addition, as for example in polynoms, Babbage chose this principle for the construction of his first calculating machines that besides a saving function only required an addition function. His plans for this "difference engine" from 1822 were a development independent of the concept designed by the German engineer Johann Helfrich Müller. A prototype, the "difference engine no 1" is based on the decimal system and was completed by Babbage in 1832. Between 1847 and 1849, he designed the "difference engine no 2" which was originally intended to have a printer, but Babbage failed to finance the project due to the immense costs for the fine mechanical workmanship required. However, between 1840 and 1854 the Swede Pehr Georg Scheutz

(1785–1873) and his son Edvard (1821–1881), on the basis of Babbage's ideas, succeeded in constructing several fully functioning difference engines, some of them also having a printing device.

Fig. 2.26 Part of the "difference engine" [4, p. 33].

Babbage's main concern was increasingly with the construction of a machine capable of solving any kind of mathematical task: "The Analytical Engine... is not merely adapted for tabulating the results of one particular function and of no other, but for developing and tabulating any function whatever. In fact the engine may be described as being the material expression of any indefinite function of any degree of generality and complexity [...]." [15, p. 109]. The "analytical engine" was to consist of the following components: a unit for arithmetical operations, a number memory with capacity for 1,000 numbers with 50 decimal places each, a control unit for calculation operations, data transfer and program flow as well as units for the input and output of data. Control is made possible with punched cards modeled on Jacquard's looms. "Operation cards" determine the type of calculation in the arithmetic unit. "Number cards" include values for numeric constants that may both be already given and also punched through a printer as a result of calculations and used directly for further calculating operations. "Directive cards" serve as data transfer between the arithmetic unit and memory by assigning certain positions in the memory to variables. Further, they consign the content of memory addresses to the arithmetic processing, whereby in this case also the respective entry may be erased in the memory. The concrete design of this "analytical engine" comprises numerous other subtypes of these punched cards that, however, may be subsumed under one of these three models. Below, there is a scheme[10] of a calculation of the expression: $(ab + c)^2$ by means of "operation cards" and "directive cards" (OC and DC, see table 2.2).

DC	OC	
1st	...	Places a on column 1 of Store
2nd	...	Places b on column 2 of Store
3rd	...	Places c on column 3 of Store
4th	...	Places d on column 4 of Store
5th	...	Brings a from Store to Mill
6th	...	Brings b from Store to Mill
.	1	Multiplies a and b = p
7th	...	Takes p to column 5 of Store where it is kept for use and record
8th	...	Brings p into Mill
9th	...	Brings c into Mill
.	2	Adds p and c = q
10th	...	Takes q to column 6 of Store
11th	...	Brings d into Mill
12th	...	Brings q into Mill
.	3	Multiplies d and q = p2
13th	...	Takes p2 to column 7 of Store
14th	...	Takes p2 to printing or stereo-moulding apparatus

Table 2.2 Scheme of a calculation by the "difference engine."

Fig. 2.27 Part of the "analytical engine" [4, p. 34].

[10] Scheme taken from [24].

The conceptual units of the "analytical engine" and its cards represent an analogue model of a processor which is connected to the central memory via data and address bus. The programming basically comprises assignments of value to certain memory addresses that are consigned to the arithmetic unit in a certain order. This concept, too, exhibits the main features of today's assembler programming. From 1833, Babbage worked on the concepts for his "analytical engine," but his project was not to be realized in his lifetime because of the enormous costs and technical effort. However, his plan was technically correct and its conceptual basics can be found in the developments of Zuse, Aiken and others.

Fig. 2.28 Ada Countess of Lovelace, contemporary engraving. Photo Herbert Matis. With kind permission of Herbert Matis.

Alongside Babbage, Ada Countess of Lovelace (1815–1852) played an important role. She was a self-educated mathematician and worked together with Babbage on the concept of the "analytical engine." In 1840, Babbage gave a course of lectures in Turin on his invention. On the basis of this material, the Italian mathematician Luigi Frederico Menabrea published the article "Notions sur la machine analytique de Charles Babbage" in 1842. Ada Lovelace translated this text to English in 1843 and added her own detailed considerations to the material, including the concepts of the *loop, subroutine* and the *conditional jump.* The Countess of Lovelace is considered to be the first female programmer in the history of informatics – the programming language Ada, developed in the 1970s, is named after her.[11]

[11] For a detailed description of the structure and the programming of the "analytical engine," see [24].

2.13 The Implementation of Logical Operations

After Leibniz' contributions to the mathematization of logic, George Boole (1815–1864) developed in his publications "The Mathematical Analysis of Logic" (1847) and "An Investigation of the Laws of Thought" (1854) a formalism for the representation of propositional logic. The works by William Hamilton (1788–1856), Augustus de Morgan (1806–1871), John Stuart Mill (1806–1873), John Venn (1834–1923), William Stanley Jevons (1835–1882) and Charles Peirce (1839–1914) must also be seen in this context. In Germany, too, Hermann Günther Grassmann (1809–1877) and his brother Robert (1815–1901) established a school for the mathematization of logic. However, these works have largely been ignored, with the exception of the works of Ernst Schröder (1841–1902) on the algebra of logic.

In "A Symbolic Analysis of Relay and Switching Circuits" (1940, MIT, Massachusetts Institute of Technology), Claude Elwood Shannon (1916–2001), later to become the founder of information theory, described the application of *Boolean algebra* in electronic circuits and laid the foundation for the construction of digital computers.

An algebra named after Boole, also referred to as *Boolean lattice*, uses the logical operators AND (conjunction, symbol: $\wedge$), OR (disjunction, symbol: $\vee$) and NOT (negation, symbol: $\neg$) and applies as elements "1" or "0," interpreted logically as true or false. Other operations such as NAND, NOR, EXOR and NEXOR can be combined by consecutively switching the operations mentioned before. In the switching algebra, this architecture enables a logical combination system that may be realized, on the basis of the elements "0" and "1," through voltage in electric circuits. In such a system, logical input variables provide value combinations that set the output variable to 0 or 1. Figure 2.29 shows the realization of a logical expression by means of a circuit.

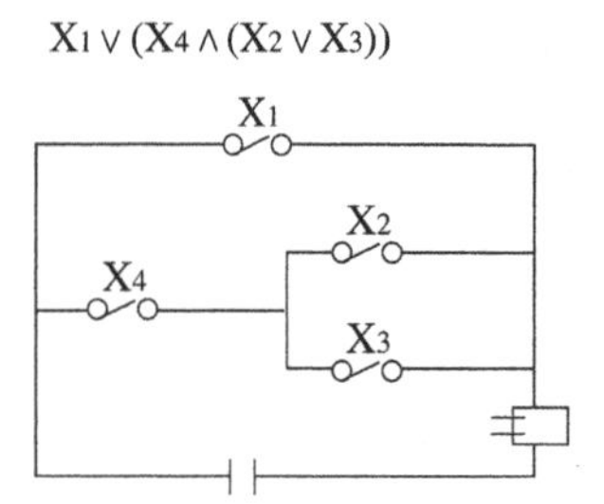

Fig. 2.29 Logical expression by means of a circuit.

Already around 1870, Stanley Jevons described in his article "On the mechanical performance of Logical Performance" a machine for the automatic processing of logical expressions. The "logical piano" is an apparatus similar to a piano, whose 21 keys are assigned with the elements of a logical operation. The left and the right half of the keyboard are assigned with the *subject* and *predicate* of a *judgment*, the middle key functions as *copula*. In this context, the subject refers to an expression about which something is said, the predicate is a characteristic that is assigned to the expression, the judgment refers to the whole statement, both as linguistic expression and to the included statement, and the copula means the connecting element between argumentation and consequence. Other keys are responsible for mechanical functions as well as *disjunctive* (mutually exclusive) *combinations*. In order to solve the expression, the keys are pressed according to the symbols of the starting premise; after getting the output premise, the logical piano shows the conclusion of the deduction.

An application of logical combinations for the generation of musical structure is represented by the theory of *sieves* applied by Iannis Xenakis (1922–2001) in a number of his compositions. In this principle, numeral arithmetic series yield new series by applying set operations. For example, with the given number series A = 0, 2, 4, 6, 8 (with step size 2) and B = 0, 3, 6, 9 (step size 3), the result of the union is: 0, 2, 3, 4, 6, 8, 9. The series can be applied in various ways for musical parameters such as pitch or duration, so that the union in the abovementioned example, if mapped on a chromatic scale, results in the notes C, D, D#, E, F#, G#, A by choosing C as a starting point. The overlapping of different periodical events can also be found in the principles of *additive rhythmic*, and is a fundamental procedure in the compositional system of Josef Schillinger (1895–1943) [21, 22].

2.14 On Formally Undecidable Propositions

Boolean algebra enables as a propositional logic the combination of statements, but disregards the scope and characteristics of representational objects. Ludwig Gottlob Frege (1848–1925) extended in his works, above all in the "Begriffsschrift" of 1879, the application range of logic with a system which is effectively the forerunner of today's *first-order logic* (also *first-order predicate calculus*). In first-order logic, *predicate symbols* and *function symbols* describe the characteristics of objects to be associated; their scope is expressed by *quantifiers*. Propositions that are combined atomically in propositional logic by operators are expressed here with *terms* and *predicates*. A term indicates the name of an object from the scope of logical operation, such as a substantive. The predicate refers to the name of characteristics, relations and classes, such as verbs, adjectives or similar. For objects unknown at the moment, variables are introduced. Symbols telling how many objects the predicate is asserted are known as *quantifiers*. The universal quantifier refers to all objects, whereas the existential quantifier refers to at least one object in the domain of discourse.

Through further development by Bertrand Arthur William Russell (1872–1970) and Alfred North Whitehead (1861–1947), the predicate calculus becomes an instrument of fundamental mathematical research, whose objectives lead to polarizations among scientists. The "formalists" under David Hilbert (1862–1943) aimed at putting basic mathematical premises on an axiomatic basis and at proving their

Fig. 2.30 Ludwig Gottlob Frege. © akg-images.

consistency. In opposition to this group were the "intuitionists," founded by Luitzen Egbertus Jan Brouwer (1881–1966), questioning the principal possibility of such an axiomatization. The objections of the intuitionists caused David Hilbert, in an epic endeavor, to put the arithmetics of natural numbers on an axiomatically confirmed basis and also to prove the consistency of mathematical domains such as set theory.

Fig. 2.31 David Hilbert. © bpk.

The position of the formalists was later strengthened by the "Principia Mathematica" (1910–1913). In this essential work, Russel and Whitehead, referring to Georg Cantor's (1845–1918) set theory as well as the works of Frege and Giuseppe Peano (1858–1932), succeeded in putting some fields of mathematics on a confirmed axiomatic basis. In 1928, Hilbert and Wilhelm Ackermann's (1896–1962) "Principles of Theoretical Logic" was published, tackling basic considerations of predicate calculus and of automatic processing of logical calculus. An interesting aspect of this work is the idea that there could be an algorithm being able to decide for any first order logical expression if its statement is true or false. This question known as the *decidability problem* may also be demonstrated by means of the *halting problem* of a Turing machine (see below). The mentioned works made the formalists' objectives seem increasingly within reach, until Kurt Gödel (1906–1978) in his publication "Über formal unentscheidbare Sätze der Principia Mathematica und verwandter Systeme" ("On Formally Undecidable Propositions of the Principia Mathematica and Related Systems") of 1931 set clear limits to their efforts. He showed first that in a sufficiently powerful system containing at least arithmetics, the consistency of the system cannot be proven within the system. Secondly, he succeeded in proving that in such a system propositions can be made that within this system are neither provable nor disprovable.

Fig. 2.32 Kurt Gödel. Kurt Gödel Papers. Manuscripts Division. Department of Rare Books and Special Collections. Princeton University Library. Reproduced with kind permission by Princeton University Library.

The works of Albert Thoralf Skolem (1887–1963) and Jacques Herbrand (1908–1931), finally led to the restriction regarding the predicate calculus that for a proposition true in the calculus, its truth can be proven in a finite number of steps; in the other case the proof may be either successful or not. This situation is also referred to as the *semi-decidability* of the predicate calculus. These findings advanced also the development of automatic proof systems, the first of such being developed in 1954 by Martin Davis at the Institute for Advanced Studies in Princeton. The "Logic Theorist" (1954) by Allen Newell, Herbert Alexander Simon and Cliff Shaw included heuristics (experience-based strategies) in the decision making process for the first time.

2.15 From Census Collector to Chess World Champion

For the development of calculating machines, the principle of punched card control developed by Basile Bouchon was for the first time applied to a data processing system by Hermann Hollerith (1860–1929).

Hollerith was a mining engineer and worked from 1879 in the American Census Agency. In 1882, he taught engineering sciences at MIT and conducted experiments on data storage on punched tapes. On the basis of an idea of the physician and health statistician John Shaw Billings, he was engaged in constructing a census machine with a control system according to the principle of Jacquard's looms. In the course of his work at the patent office in Washington from 1883 on, he secured the rights for his invention. From 1889 on, the procedure was successfully tested in Baltimore and New Jersey and in 1890 was first used for census taking in America, as well as in Austria.

Fig. 2.33 Hollerith's punched card tabulator. With kind permission of IBM, Corporate Archives, Somers, NY.

In 1893, Hollerith founded the Tabulating Machine Company which merged in 1911 with two other companies, and was renamed as the Computing Tabulating Recording Company.

Fig. 2.34 Tabulating Machine Factory in 1893. With kind permission of IBM Corporate Archives, Somers, NY.

In 1924, with Thomas Watson as chairman, the company again changed its name, to become the International Business Machines Corporation (IBM).

Fig. 2.35 Thomas Watson. With kind permission of IBM Corporate Archives, Somers, NY.

Between 1923 and 1927, the first analogue computer for solving differential equations was developed under Vannevar Bush (1890–1974) at MIT. Analogue computers represent their calculations on the basis of seamless state transitions and the results can be directly read off a measurement device. However, measuring analogue values causes fault tolerances that influence the accuracy of the result. In contrast to that, digital computers calculate on the basis of discrete number representations. The first digital computers in this sense were built beginning in the 1930s in Germany by the Berliner construction engineer Konrad Zuse (1910–1995), and in the United States by Howard Hathaway Aiken (1900–1972). These computers served as a model for all following developments and may already be accurately referred to as "computers" in today's sense.

Fig. 2.36 Konrad Zuse and Howard Hathaway Aiken [4, p. 84, 67].

Etymologically, the term derives from the Latin "computare," French "computer" and English "to compute." According to the "Barnhart Concise Dictionary of Etymology," in the English speaking world the term "computer" was used to refer to a person from 1646 on. From 1897, it may also have been used for a calculating machine. Beginning in 1945, the term was introduced in the publications of Eckert, Neumann and others on the computer systems EDVAC (Electronic Delay Storage Automatic Computer) and ENIAC and is since its definition in the "Oxford English Dictionary" of 1946 (where also different types are distinguished), a common designation for calculating machines.

From 1934 to 1938, Konrad Zuse developed his first model of a program-based computing machine. The Z1 (Zuse 1) was made entirely of mechanical components. Zuse used a memory capable of storing 16 binary numbers of 24 bits, each consisting of re-locatable metal sheets. Although his concept was thought out correctly, some problems arose due to the mechanical material load.

In the follow-up models Z2 (1940) and Z3 (1941), the mechanical switching elements were replaced by relays. The Z3, considered to be the first programmable electronic computer, enabled the processing of floating-point numbers not implemented in similar models, such as Mark I, ENIAC and others, of that time. The half-logarithmic notation used by Zuse divides the number into exponent base 2 and

Fig. 2.37 Zuse's Z1 [4, p. 85].

mantissa and so allows for a more efficient number representation than the fixed-point notation used in similar computer models. The basic structure of the Z3 comprises the elements shown in figure 2.38: The arithmetic unit is separated from the

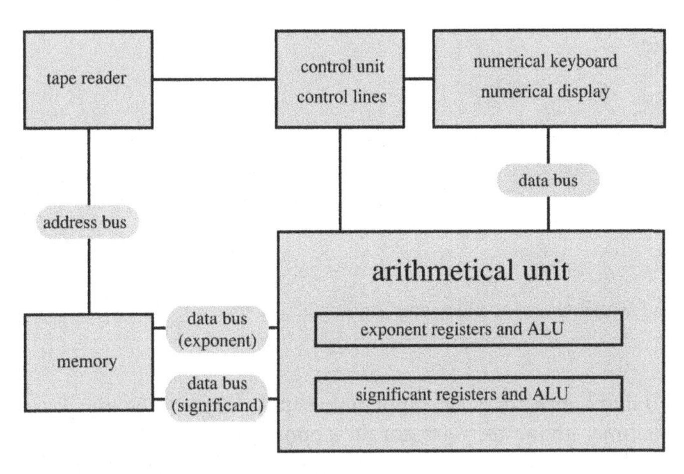

Fig. 2.38 Block diagram of the Z3.

memory. Input and output of numbers happen with decimal numbers, the intern processing is done in binary numbers. The binary memory has a memory capacity of 64 floating-point numbers. Memory and arithmetic unit are linked by a data bus. Every number is represented in the memory by three binary fields. The first bit stores the algebraic sign; the next seven bits store the exponent and the last 14 bits contain the value of the mantissa. As result of these specifications, the smallest possible representable number is 2^{-63} and the largest $1.9999 * 2^{62}$. The arithmetic unit contains two registers each for the calculation of the exponent and the mantissa. The

Fig. 2.39 Details of the mechanical memory of the Z-series [30, p. 172].

following nine commands (table 2.3) from input and output storage and arithmetic calculation are stored on a punched tape with lines of eight bit each and may be combined in any order.

Lu	read keyboard
Ld	display result
P r z	read address z
P s z	store in address z
Lm	multiplication
Li	division
Lw	square root
Ls1	addition
Ls2	subtraction

Table 2.3 Input, output and arithmetic calculation commands.

A numeral keyboard and lamps serve for input and output. The commands "Lu" and "Ld" pause the machine, giving the operator the opportunity to type in a number or read a result, after which the machine continues. Before each calculation, the values for the registers "R1" and "R2" must be entered through the keyboard before they are loaded into the registers. "R1" and "R2" may also be loaded through values from memory. Sequencing is done by a clocked control unit running through five-stage cycles. Stages IV and V transmit information and I, II and III carry out arithmetic operations with the loaded values. So, a command consists of the determination of arguments, the calculation procedure and the rewriting of the result. For the arguments, the values in "R1" and "R2" are used; the result is then stored in "R1." Zuse applied a procedure already used by Babbage, the computing plan for solving complex formulas in elementary operations. The following example shows a possible computing plan [30, p. 169]:

$\sqrt{a^2+b^2} = c$; $a = V_1$; $b = V_2$; $c = V_3$; $V_1 \cdot V_1 = V_4$; $V_2 \cdot V_2 = V_5$; $V_4 + V_5 = V_6$; $\sqrt{V_6} = V_3$.

A function of the type $((a_4 \cdot x + a_3) \cdot x + a_2) \cdot x + a_1$ would be processed in Z1 and Z3, as can be seen in table 2.4.[12]

Pr 4	load a4 in R1
Pr 5	load x in R2
Lm	multiply R1 and R2, result in R1
Pr 3	load a3 in R2
Ls1	add R1 and R2, result in R1
Pr 5	load x in R2
Lm	multiply R1 and R2, result in R1
Pr 2	load a2 in R2
Ls1	add R1 and R2, result in R1
Pr 5	load x in R2
Lm	multiply R1 and R2, result in R1
Pr 1	load a1 in R2
Ls1	add R1 and R2, result in R1
Ld	display result

Table 2.4 Processing of a computing plan.

This kind of programming enables, for example, iterative processes, since the punched tape may run in a loop. However, branch instructions that Zuse later implements in his Z4, developed between 1942 and 1944, are missing here.

Fig. 2.40 Reconstruction of Z3 in the Deutsches Museum in Munich [30, p. 58].

[12] For a detailed description of the architecture of the Z1 and Z3, see [18].

Parallel to the development of Zuse's computers, in the USA significant new developments were also occurring. From 1934 on, Howard Hathaway Aiken (1900–1973) designed, in cooperation with IBM, the Mark I, first presented to the public at Harvard in 1944. Aiken's machine is based on the decimal system and consists of an interconnection of Hollerith's machines for the calculation of general arithmetic tasks. At this time, the costs for computer systems were enormous, meaning that besides IBM also the US Navy helped finance this project; the system was consequently used mainly for calculations of ballistic tasks. Between 1945 and 1952 the follow-up models Mark II to Mark IV, in which besides the relays already the significantly faster electron tubes were applied as switching elements, were developed.

Fig. 2.41 Harvard Mark I. With kind permission of IBM, Corporate Archives, Somers, NY.

Grace Murray Hopper (1906–1992) worked on the programming of Mark I, designing the concept of *subroutine* and contributing significantly to the development of the first compiler, which was finished in 1952 as a military invention. Grace Hopper was alongside Ada Lovelace one of the female pioneers of computer development and after 1945 was promoted to admiral to become the highest ranking woman of the US Army.

Between 1937 and 1942, a binary computer for solving equation systems was developed by John Vincent Atanasoff (1903–1995) and his doctoral candidate Clifford Edward Berry (1918–1963) at Iowa State College. The Atanasoff–Berry Computer (short: ABC) completely abandoned complicated mechanical relay technology in favor of the new electron tubes and disposed of a dynamic memory as well as independent units of calculation.

The ENIAC I, realized by the Moore School of Electrical Engineering at the University of Pennsylvania in cooperation with the US Army, was the prototype of the modern computer which exclusively consists of electronic components, except for the periphery devices. This large-capacity computer was developed at the Moore School under the direction of electronics technician John Presper Eckert (1919–1995), the physicist John William Mauchly (1907–1980) with the participation of the mathematician Herman Goldstine from the Army. The ENIAC I was presented

Fig. 2.42 Grace Murray Hopper. Official U.S. Navy Photograph, from the collections of the Naval Historical Center.

Fig. 2.43 John Presper Eckert and John William Mauchly [4, p. 111].

to the public in 1946, operating until 1955 and constantly undergoing improvement during its operation time. From 1944 on, the development team of the ENIAC I was already planning the next computer, the EDVAC (Electronic Delay Storage Automatic Computer). In contrast to the ENIAC I, which is based on the decimal system, the data representation is binary. Loops and jump commands were possible and later, based on a concept of John von Neumann (1903–1957), it became possible to store program commands besides data in the memory.

At this time, Neumann worked under the direction of Robert Oppenheimer (1904–1967) in the Los Alamos National Laboratory on mathematical problems for the construction of an atomic bomb. In 1945, he wrote the "First draft of a report on the EDVAC" for the planned computer, in which he presented the concept of a digital universal computer programmable from memory. This draft also lends its name to the so-called *Von Neumann architecture* that will be explained later in more detail. The EDVAC was finished only in 1952, although the architecture pro-

Fig. 2.44 "ENIAC." U. S. Army Photo.

posed by Neumann was already implemented by 1945 on the IAS computer of the Princeton Institute of Advanced Studies, operating until 1960. Different versions of the IAS computer were built; one of them is the ILLIAC at the University of Illinois, on which also the first computer generated composition was produced by Lejaren Hiller and Leonard Isaacson in 1956 (see chapters 3 and 10 for details).

The replacement of the large and fault-prone electron tubes was already in sight by 1948, when William Bradford Schockley, John Ardeen and Walter Hauser Brittain filed the patent for the transistor. The first fully transistorized computer TRADIC was built only six years later by Bell Laboratories. In 1949, Maurice Vincent Wilkes at the University of Cambridge built the EDSAC (Electronic Delay Storage Automatic Computer), a computer programmable from memory which implemented von Neumann architecture.

At Texas Instruments in 1958, Jack St. Clair Kilby developed the integrated circuit (IC). Robert N. Noyce of the rival business Fairchild filed a patent for his version of an IC shortly after Kilby. The lawsuit between Texas Instruments and Fairchild ended with a settlement. With another employee of Fairchild, Gordon E. Moore founded the Integrated Electronics (Intel) in 1968, in 1971 developing the 4004, one of the first micro processors. In 1973, Rank Xerox developed the first personal computer, the Alto. Already by 1974, the Altair 8800, the first assembly kit of a PC for less than 400 USD, was being offered by MITS (Micro Instrumentation and Telemetry Systems), a one-man company run by electronics engineer Edward Roberts. The sales exceeded all expectations and the Altair was adapted by the users in various ways.

The programmer Paul Allen and his friend, the student Bill Gates created a version of the programming language Basic for the Altair, up to then only programmable in machine-code, and started their own company Microsoft in 1975. In 1976, Stephen Paul Jobs and Stephen Wozniac built their own computer due to lack

Fig. 2.45 Robert Oppenheimer and John von Neumann at the Institute of Advanced Study. Alan Richards photographer. Courtesy of the Archives, Institute for Advanced Study. Princeton, NJ, USA. With kind permission.

of money and finally began to offer their Apple I as a kit for 666 USD. In the following years, other very successful computers for private use were created by the company Atari (models 400 and 800 in 1979), Commodore (VC 20 in 1980) as well as from 1980 on IBM PCs with the MS-DOS operation system of Microsoft. But it was not only the development of micro computers that was advanced enormously. By 1964, Seymour Roger Cray with the company Control Data Corporation developed the CDC 6600, a first model of a new generation of "super computers." Among these, Deep Blue created by IBM reached great popularity in 1997 by defeating the World Champion Garry Kasparov in a chess tournament.

The improvements in this field mentioned in this chapter only serve as examples; in the Soviet Union during the Cold War, computer technology was flourishing with pioneers such as Sergey Alexeyevitch Lebedev (1902–1974) and Viktor Mikhailovich Glushkov (1923–1982).[13]

[13] For a detailed overview see [15] and [27].

Fig. 2.46 Deep Blue. With kind permission of IBM Corporate Archives, Somers, NY.

2.16 Automata and Computability

For the designing of computers, a number of questions regarding programming and architecture as well as the potential of automated computing arise from the context of the technical achievements of that time. In 1936, the British mathematician Alan Mathison Turing (1912–1954) developed a mathematical model of a machine to solve Gödel's Incompleteness Theorem. This theorem makes the generality of a statement dependant on its formal provability. The question of provability depends on the possibility of compatibility. Turing's automaton model (*Turing machine*) defines a class of compatible functions whose way of processing serves as a suggestion for the development of modern imperative programming languages. Alonzo Church (1903–1995) and Stephen Cole Kleene (1909–1994) invented the *lambda calculus*, an equivalent procedure that greatly influenced the development of programming languages. In *imperative* or *procedural* programming languages, computations are described as sequences of commands and are generally performed consecutively. *Functional* programming languages conceive computation as the evaluation of mathematical functions and do not use a determined order of computations. In other words, within a functional environment the input data is processed with functions in order to receive the output data.

Provided that it is principally computable, any algorithm may be processed also by the model of a Turing machine. The Turing machine is a special example of an automaton. The term "automaton" derives etymologically from old Greek as well as Indo-Germanic, meaning "acting of one's own will." An automaton can be defined as an abstract structure in a specific environment comprising everything but the automaton itself. The automaton is in a possible *state* and can move to another state by reacting to environmental stimuli. The kind of *state transition* depends on the current state and the character of the stimulus. The states and stimuli may be thought of as discrete, clearly distinguishable events, the stimuli occurring in a clocked cycle. At the beginning, the automaton is in the initial state and moves to another state by reacting to stimuli from its environment. If within the next unit of time no new input is presented, this is considered as an input, leading the automaton to transition in itself. An abstract symbol is set for each stimulus; if the state does not change, the assumed symbol is a blank field. In order to be able to recognize the symbols, the automaton is equipped with an abstract sensor or *reading head*. It is assumed that the automaton can only react with a finite set of states that it moves to depending on the different symbols. These symbols for the stimulation of the automaton together form a conceptually coherent set referred to as the *alphabet*. A symbol string consisting of elements from the alphabet represents the input of the automaton. In the same way a state of the automaton is taken as its initial state, another or several other states are interpreted as *accepting* or *final* states. In case its environment may move the automaton in its initial state anytime, and reaching a final state is taken as an output, this machine is also referred to as a *recognizer*. However, if not a final state but the state transitions of a machine are interpreted as output, this system is known as a *Moore machine*. If the automaton is capable of generating an output symbol based on its current state in combination with an input, this model is referred to as a *Mealy machine*. The following applies to all of the mentioned automata:

- A finite alphabet of input symbols.
- A finite set of possible states with one initial state and at least one final state.
- Discrete operation: Only symbols of the alphabet as well as the possible states are considered, transitions are not taken into consideration.
- Sequential operation: Input, state changes and output happen sequentially without having the possibility of reversing or jumping over operation steps.
- Deterministic operation: Each possible reaction of the automaton is clearly determined by the current input symbol as well as the current state.

In case an automaton meets these criteria, it is referred to as a *finite automaton* that may also be applied to formulate rewriting rules in algorithmic composition (see chapter 4).

A Turing machine consists basically of a tape serving as the interface to the environment, which is divided into a finite number of cells, any cell of which may contain a symbol from a finite alphabet. The automaton comprises a device for reading, erasing and writing symbols in the current cell. This abstract unit called *head* can also move the tape from the current cell left and right, but only one cell at a time.

- The automaton is in a particular state, the input symbol is read.
- Input symbol and state determine the *argument*. The argument determines the next output symbol that is written on the tape after the input symbol has been erased.
- The head moves left or right, depending on the instructions of the argument.

- The automaton moves to the next state determined by the argument.

After reaching one of its final states, the automaton halts and indicates this in some way or another. In case the head moves beyond the end of the tape, another blank symbol is added in order to make the automaton halt. A common convention of describing Turing machines is the argument quintuple, its first two entries representing current state and input symbol and the other three entries the output symbol, the moving direction of the head and the next state.

The following program (table 2.5)[14] enables the identification of palindromes and uses the following notation: 1. current state, 2. current symbol, 3. new state, 4. symbol to be written, 5. moving direction. If the input consisting of the symbols "A" and "B" is identified as a palindrome, the output is to be "YES"; and if this is not the case, "NO." In both cases the machine holds after the last output. The "O" is an optional symbol to mark the beginning of the tape; "_" indicates a blank field.

(1 _ 2 O R)			
(2 A 3 _ R)	(2 B 4 _ R)	(2 _ 7 _ L)	
(3 A 3 A R)	(3 B 3 B R)	(3 _ 5 _ L)	
(4 A 4 A R)	(4 B 4 B R)	(4 _ 6 _ L)	
(5 A 11 _ L)	(5 B 12 _ L)	(5 _ 7 _ L)	
(6 A 12 _ L)	(6 B 11 _ L)	(6 _ 7 _ L)	
(7 _ 7 _ L)	(7 O 8 _ R)		
(8 _ 9 Y R)			
(9 _ 10 E R)			
(10 _ H S R)			
(11 A 11 A L)	(11 B 11 B L)	(11 _ 2 _ R)	
(12 A 12 _ L)	(12 B 12 _ L)	(12 _ 12 _ L)	(12 O 13 _ R)
(13 _ 14 N R)			
(14 _ H O R)			

Table 2.5 Turing program for the identification of palindromes.

In case the input consists of the symbol string "ABA," the program lines of the left column are applied. The current state of the tape can be seen in the right column, the apostrophe showing the position of the head (table 2.6). The program for detecting palindromes uses the strategy of repeatedly erasing identical symbols at the beginning and at the end of the symbol string, whereas the symbols in the middle are replaced by themselves.

By dividing complex computing processes in basic operations, the Turing machine is at the same time a model for computation in general. The halting problem of the Turing machine, which is related to the decision problem and Gödel's incompleteness theorem, raises the question of whether a general procedure exists that may decide for any algorithm if it terminates or not. This halting problem is unsolvable, so for an arbitrary algorithm it cannot be said from the outset if the Turing machine will halt in a finite number of steps or not.

[14] Taken in a modified form from [6].

(1 _ 2 O R)	O A' B A _
(2 A 3 _ R)	O _ B' A _
(3 B 3 B R)	O _ B A' _
(3 A 3 A R)	O _ B A _'
(3 _ 5 _ L)	O _ B A' _
(5 A 11 _ L)	O _ B' _ _
(11 B 11 B L)	O _' B _ _
(11 _ 2 _ R)	O _ B' _ _
(2 B 4 _ R)	O _ _ _' _
(4 _ 6 _ L)	O _ _' _ _
(6 _ 7 _ L)	O _' _ _ _
(7 _ 7 _ L)	O' _ _ _ _
(7 O 8 _ R)	_' _ _ _ _
(8 _ 9 Y R)	_ Y _ _ _
(9 _ 10 E R)	_ Y E _ _
(10 _ H S R)	_ Y E S _

Table 2.6 Example transitions of the automaton at a given input.

2.17 The Model of a Universal Computer

The Turing machine developed at this time is an abstract model for computability, whereas the von Neumann architecture mentioned before provides a concrete construction plan for the creation of computer systems. The concept presented in "First draft of a report on the EDVAC" comprises the following components: *Control unit*, *arithmetic unit*, *memory*, as well as *input* and *output devices*.

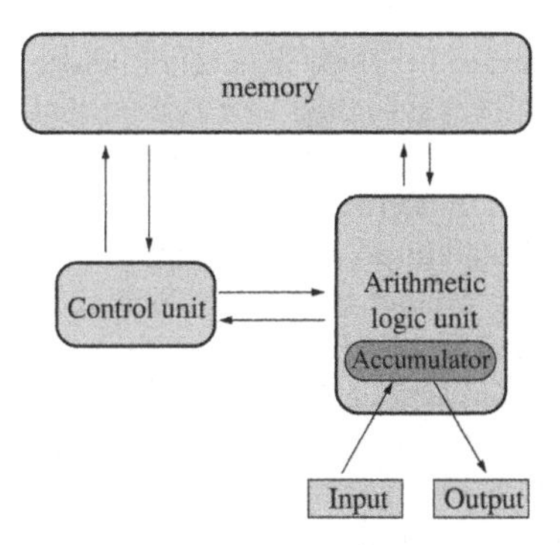

Fig. 2.47 Scheme of a von Neumann computer.

The following basic conditions apply:

- The structure of the computer is universal, independent of the problem to be solved.
- The memory contains data, program, intermediate and end results.

- The memory is divided in consecutively numbered cells of the same size, the number of the cell being the address for access from programs, both for retrieval and change of its content.
- The program consists of a sequence of commands that must be processed sequentially; however, a jump command may be used which allows branching.
- Basic commands comprise arithmetic, logical as well as transport commands controlling the data transfer between input and output, arithmetic unit, memory, etc.
- The data is encoded in a binary system.

The arithmetic unit consists of the *accumulator*, a unit for storing binary information of particular lengths as well as an *arithmetic unit* and a *sequential control*. Together these units form the *arithmetic logic unit* (*ALU*) that computes values assigned by the control unit and returns the results. The *control unit* is responsible for the sequential control within the computer and controls the data transfer between the components.

2.18 Programming

The development of computers also requires strategies for the input of data and programs. Between 1939 und 1945, Konrad Zuse developed the Plankalkül,[15] which allowed for the description of a computer program in an abstract language, as a first concept of a higher programming language. The Plankalkül represents a formalism of first-order logic with added imperative constructs. Zuse was challenged by the idea of investigating its universal applicability. He dealt with the possible implementation of chess strategies and generated a program in 1945, five years before Shannon's works on the same subject. The Plankalkül is designed as a system that is able to operate with general logical expressions. Every Plankalkül-program may be represented as a function with variable arguments; these functions can be called from numerous positions of the program. The Plankalkül as a first example of a higher programming language shows many parallels to other languages developed considerably later; however, it was only implemented for the first time in the year 2000 by the Department of Informatics at the Freie Universität Berlin. [19].

A higher programming language (also: third generation programming language) allows an extensively machine-independent notation of algorithms, whereas *machine languages* (also: first generation programming languages) use processor-specific number codes and *assembly languages* (also: second generation programming languages) use symbolically encoded number codes for program instructions. *Applicative programming languages* (also: fourth generation programming languages) extend the concept of higher programming languages for solving application-specific problems. Fifth generation languages, such as Prolog (from 1972 on), which

[15] For a description of the basic structure of the Plankalkül and an application for solving chess problems, see [30, p. 190ff].

is based on the predicate calculus, enable the input of a problem without having to implement an explicit routine or algorithm to solve it.

The classification of programming languages in generations does not represent a hierarchical order in today's understanding. Early languages of third generations such as Fortran (from 1955 on) and Lisp (from 1958 on) are, in modified forms, still in use today.[16] Depending on the structure, programming languages may be divided into different concepts. In procedural languages such as Fortran, Basic (from 1964 on) and Pascal" (from 1970 on), instructions are in general processed consecutively. In functional languages such as Lisp, functions are defined that may be called independent of their position within the program, also within other functions. Object-oriented languages such as Smalltalk (from 1972 on) define objects representing autonomous units of data and algorithms. Objects that share the same internal structure are said to belong to the same *class*. The basic characteristics of a class may be transferred hierarchically to other classes by the concept of *heredity*. The behaviors of the objects as well as their mutual relations are described with so-called *methods*. Together with machine languages, assembler and rule-based languages such as Prolog, these paradigms form the basic models of programming languages. Visual programming environments, in which in addition to the actual code visual objects may be manipulated on the screen, have been developed for programming languages of different types. Current programming languages as well as advancements of existing languages mostly unify different paradigms in their functional range.

2.19 The Computer in Algorithmic Composition

These developments of algorithmic thinking processes as well as physical instruments for their processing are the basis for the processing of algorithms generating musical structure. The first completely computer-generated composition on a symbolic level was produced by Lejaren Hiller and Leonard Isaacson from 1955 to 1956 with the "Illiac Suite" on the ILLIAC computer at the University of Illinois.[17] The symbolic level refers to the fact that the output of the system represents note values that must be interpreted by the musician.

Along with the development of higher programming languages, the first computer music systems for algorithmic composition were also generated. Their concepts are based either on general programming languages or have been designed especially for the particular application. In 1963, Hiller and Robert Baker developed Musicomp, the first computer-assisted composition environment. Max Mathews' and Richard F. Moore's system Groove marked the beginning of the developments in the 1970s. Midi Lisp, Patch Work and Bol Processor were the new programming environments of the 1980s. Common Music, Symbolic Composer, Open Music and many other systems followed at the beginning of the 1990s. A number of these sys-

[16] In algorithmic composition, a number of different systems are still applied today that are based on Lisp or Scheme.

[17] For a detailed description of this project see [9].

Fig. 2.48 Lejaren Hiller. © University Archives, University of Illinois.

tems are still in use today and are constantly being advanced. Computer music languages that were not primarily designed for algorithmic composition but may well be used for this purpose, depending on the program architecture, must also be mentioned in this context. Beginning with the languages of the MusicN family of Mathews from the 1960s on, Barry Vercoe's Csound (from 1986 on), Bill Schottstaedt's Common Lisp Music (from 1991 on) up to current developments such as PureData (PD) and SuperCollider, there is a wide range of powerful and flexible computer music systems.

The "Illiac Suite" marked the beginning of computer-assisted algorithmic composition whose advancement was influenced by other compositional tendencies as well as developments in other fields of art. In 1955, Hiller and Isaacson were working on the "Illiac Suite"; in the same year, "Metastasis" by Iannis Xenakis had its world premiere at the Donaueschinger Musiktage. This composition is characterized by a complex web of glissandi moving at different speeds: "If glissandi are long and sufficiently interlaced, we obtain sonic spaces of continuous evolution. It is possible to produce ruled surfaces by drawing the glissandi as straight lines [...] Several years later, when the architect Le Corbusier, whose collaborator I was, asked me to suggest a design for the architecture of the Philips Pavilion in Brussels, my inspiration was pin-pointed by the experiment with Metastasis." [28, p. 10]. The Philips Pavilion at the Brussels World's Fair of 1958 came into being as a synthesis of different art forms; with Xenakis' and Le Corbusier's architectural conception, Philippe Agostini's experimental film work, Xenakis' 2-track composition "Concret PH" and Edgar Varèse's seminal "Poème Electronique," which was realized in the pavilion with an elaborate multi-channel sound system.[18]

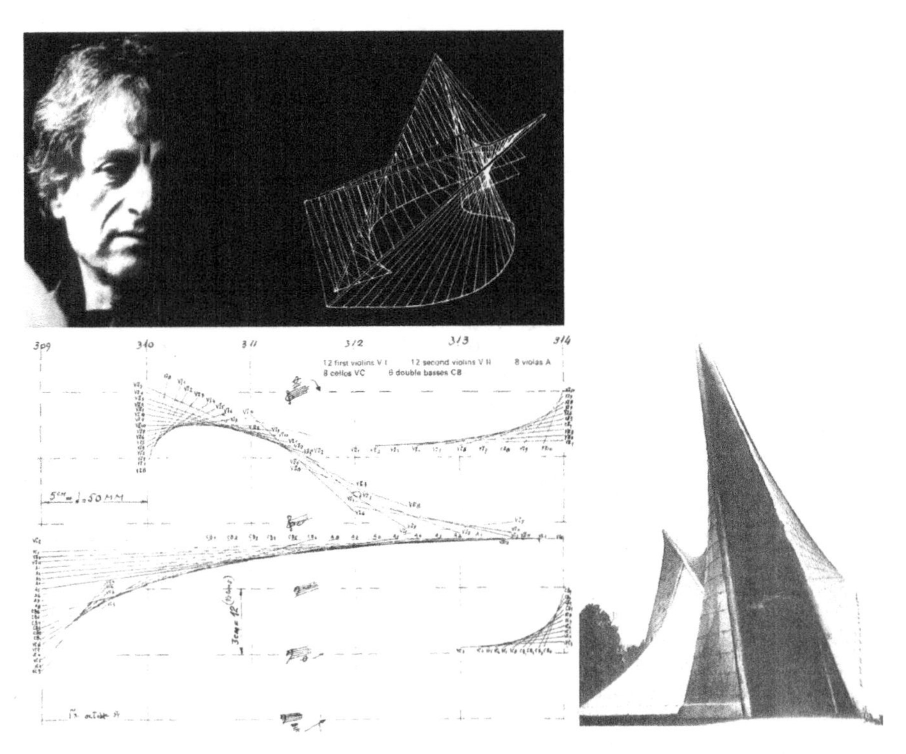

Fig. 2.49 Iannis Xenakis and his first model of the Philips Pavilion (top); string glissandi, bars 309–14 of Metastasis and the Philips Pavilion at the Brussels World's Fair (bottom) [28, p. 10, 11, 3 and from the backcover]. Reproduced with kind permission by Pendragon Press.

References

1. Bonner A (ed) (1985) Doctor Illuminatus: a Ramon Llull reader. Princeton University Press, Princeton, NJ. ISBN 0691000913
2. Borges JL (2000) The library of Babel. David R. Godine Publisher, Boston. ISBN 978-1567921236
3. Dantzig T (1968) Number: the language of science, 4th edn. Free Press, New York. ISBN 0029069904

[18] The "Poème Electronique" has been reconstructed in a virtual acoustic Philips pavilion by the Institute of Electronic Music and Acoustics (IEM) at the University of Music and Dramatic Arts Graz, Austria [29], and also by a team from the universities Virtual Reality And Multi Media Park (Torino, Italy), University of Bath (England), Technische Universität Berlin (Germany), and Silesian University of Technology (Gliwice, Poland).

4. De Beauclair W (2005) Rechnen mit Maschinen. Eine Bildgeschichte der Rechentechnik. Springer, Berlin. ISBN 3-540-24179-5
5. Eco U (1995) The search for the perfect language. translated by James Fentress. Blackwell, Oxford
6. Fachhochschule Wiesbaden – Fachbereich 06 – Informatik (2004) Turing Machine Simulator. http://wap03.informatik.fh-wiesbaden.de/weber1/turing/ Cited 12 Oct 2004
7. Glaser A (1971) History of binary and other nondecimal numeration. Anton Glaser, Southhampton. ISBN 0-938228-00-5
8. Glashoff K (2003) Gottfried Wilhelm Leibniz – die Utopie der Denkmaschine. http://www.logic.glashoff.net/Texte/GottfriedWilhelmLeibniz6.pdf Cited 27 Sep 2004
9. Hiller L, Isaacson L (1958) Musical composition with a high-speed digital computer. Journal of the Audio Engineering Society, 1958. In: Schwanauer SM, Levitt DA (eds.)(1993) Machine models of music. The MIT Press, Massachusetts. ISBN 0-262-19319-1
10. Hedges SA (1987) A dice music in the eighteenth century. Music and Letters, 59/2, 1987
11. Heisenberg W (1927) Physical principles of Quantum Theory. Zeitschrift für Physik 43, 1927, C.Eickort & F.C.Hoyt (Trans), Dover (1930)
12. Hochstetter E, Greve H-J (eds) (1966) Herrn von Leibniz Rechnung mit Null und Einz. Siemens Aktiengesellschaft, Berlin
13. Kaplan R (1999) The nothing that is. A natural history of zero. Allen Lane, London
14. Lasswitz K (1997) Die Universalbibliothek. Wehrhahn-Verlag, Laatzen. ISBN 3932324994
15. Matis H (2002) Die Wundermaschine: die unendliche Geschichte der Datenverarbeitung; von der Rechenuhr zum Internet. Ueberreuter, Wien. ISBN 3-8323-0936-5
16. Remnant P, Bennett J (eds) (1996) New essays on human understanding. Cambridge University Press, Cambridge
17. Rojas R (2000) Z1, Z2, Z3 and Z4. http://www.zib.d e/zuse/index.html Cited 12 Oct 2004
18. Rojas R(2000) Die Architektur der Rechenmaschinen Z1 und Z3. http://www.zib.de/zuse/Inhalt/Kommentare/Html/0687/0687.html Cited 9 Sep 2004
19. Rojas R, Göktekin C, Friedland G, Krüger M, Scharf L (2000) Konrad Zuses Plankalkül – Seine Genese und eine moderne Implementierung. http://www.zib.de/zuse/Inhalt/Programme/Plankalkuel/Genese/Genese.pdf Cited 13 Oct 2004
20. Sadie S (ed) (1995) The New Grove dictionary of music and musicians, 7. Macmillan Publishers, New York & London. ISBN 1-56159-174-2
21. Schillinger J (1973) Schillinger System of musical composition. Da Capo Press, New York. ISBN 0306775522
22. Schillinger J (1976) The mathematical basis of the arts. Kluwer Academic/Plenum Publishers, New York. ISBN 0306707810
23. Seife C (2000) The biography of a dangerous idea. Viking Penguin, New York
24. Walker J (2004) The Analytical Engine. The first computer. http://www.fourmilab.ch/babbage/ Cited 7 Oct 2004
25. Wiener O, Bonik M, Hödicke R (1998) Eine elementare Einführung in die Theorie der Turing Maschinen. Springer, Wien New York. ISBN 3-211-82769-2
26. Wilhelm R (1982) I Ging. Das Buch der Wandlungen. Eugen Diederichs Verlag, Köln. ISBN 3424000612
27. Williams M (1985) A history of computing technology. Prentice-Hall, Englewood Cliffs, NJ. ISBN 0-13-389917-9 01
28. Xenakis I (1992) Formalized music. Thought and mathematics in music, revised edition. Pendragon, Stuyvesant, NY. ISBN 0-945193-24-6
29. Zouhar V, Lorenz R, Musil T, Zmölnig J, Höldrich R (2005) Hearing Varèse's Poème Electronique inside a Virtual Philips Pavillion. In: Proceedings of ICAD 05-Eleventh Meeting of the International Conference on Auditory Display, Limerick, Ireland, July 6-9, 2005
30. Zuse K (1984) Der Computer – Mein Lebenswerk, 2nd edn. Springer, Berlin. ISBN 3-540-16736-6

Chapter 3
Markov Models

Markov models were introduced by the Russian mathematician Andrey Andreyevich Markov (1856–1922). Markov was a student of Pafnuty Chebyshev and worked amongst others in the field of number theory, analysis, and probability theory. From 1906 on, Markov published his first works on time dependent random variables.

Fig. 3.1 Andrey Andreyevich Markov in 1886. Contemporary engraving.

In the context of urn experiments, observations on time dependent random variables were already made by Pierre Simon de Laplace and Daniel Bernoulli [2, p. 10ff]. The first application of this method was an extensive text analysis by Markov: "In 1913 [...] Markov had the third edition of his textbook[1] published. [...] In that edition he writes, 'Let us finish the article and the whole book with a good example of dependent trials, which can be regarded approximately as a simple chain'. In what has now become the famous first application of Markov chains, A. A. Markov studied the sequence of 20,000 letters in A. S. Pushkin's poem 'Eugeny Onegin', discovering that the stationary vowel probability is $p = 0.432$, that the probability of a vowel following a vowel is $p1 = 0.128$, and that the probability of a vowel following a consonant is $p2 = 0.663$ [...]." [2, p. 16]. The term "Markov chain" (in

[1] Basharin et al. are referring to Markov's paper "Ischislenie veroyatnostej."

this work, also referred to as "Markov model") for this class of stochastic procedures was first used in 1926 in a publication of the Russian mathematician Sergey Natanovich Bernstein.

3.1 Theoretical Basis

The field of stochastics comprises probability calculus and statistics. *Stochastic processes* are used to describe a sequence of random events dependent on the time parameter (t). The set of events is called "*state space*," while the set of parameters is known as the "*parameter space*." If a stochastic process consists of a countable number of states, then it may also be referred to as a stochastic chain. In a stochastic chain, every discrete time t has a random variable X. In a *Markov chain*, it being a special kind of stochastic chain, the probability of the future state X_{t+1} (the random variable X at the time $t+1$) depends on the current state X_t. For the given times t_m and t_{m+1}, this probability is:

$$P(X_{tm+1} = j \mid X_{tm} = i) = p_{ij}(t_m, t_{m+1})$$

This expression indicates the *transition probability* of the state $X_{tm} = i$ at a given time t_m to the state $X_{tm+1} = j$ [3, p. 768].

A Markov chain can be represented by a state *transition graph*, or by a *transition matrix*, as can be seen in the example of a weather forecast in figure 3.2. The description of the edges in figure 3.2a shows the transition probabilities that can be found in transition matrix P represented as a table (3.2b). The sum of all transition probabilities in each state must equal 1. Starting from a particular state, the probabilities for a future state can be determined. These probabilities are calculated with the formula $p(t+k) = p(t) * P^k$; $p(t)$ representing the initial state, k the number of state transitions and P the transition matrix [3, p. 788].

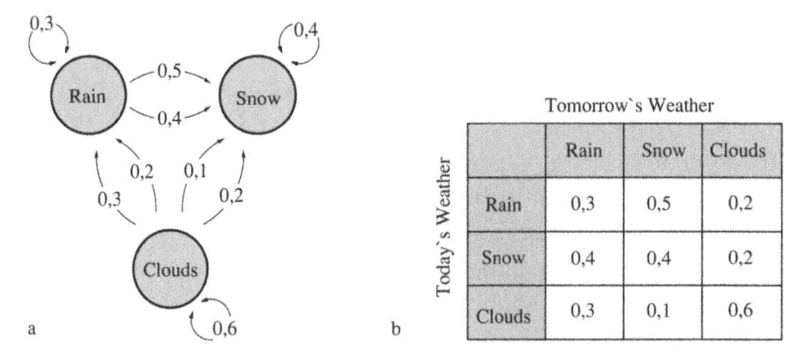

	Rain	Snow	Clouds
Rain	0,3	0,5	0,2
Snow	0,4	0,4	0,2
Clouds	0,3	0,1	0,6

Fig. 3.2 Representations of a first-order Markov chain.

If more than one past event is used in the calculation of the transition probabilities, then this is called a *higher-order Markov process*, the order indicating the number of past events that are relevant to the transition probabilities. Consequently, in a Markov model based on the tone pitches of a melodic corpus, the output sequence will more and more approach the structure of the corpus with increasing Markov orders.

If a particular sequence of length n does not occur in the corpus, then this sequence will not appear in the produced musical material in the corresponding Markov analysis and generation of nth order either. A possible solution to this problem is offered by so called "*smoothed n-grams*"[2] that use lower-order transition probabilities for the generation of higher-order transition probabilities. In this procedure, the missing transition probabilities for insufficient sequences of nth order can be acquired by interpolation with lower orders $n-1$, $n-2$, etc. At a given symbol at the current point, the transition probabilities calculated from the corpus get different weights for preceding sequences of different length. In general, Markov models can only occupy a finite number of states and can therefore be represented by finite automata[3] and within a graph representation. This possibility is especially interesting in cases where not all fields of the transition matrix need to be occupied [5, 11].

3.2 Hidden Markov Models

In hidden Narkov models (HMM), the sequences of the *observable output* symbols of a Markov model are visible, but their internal states and state transitions are not. In this case, the states of the "hidden" Markov models produce so-called *emission probabilities*, generating the musical segments (usually notes with tone pitches and lengths) as observable outputs. In the following, an HMM is explained through the example of a weather forecast. A news agency receives its information on political events in a foreign country from a correspondent at different times of the day. In his work, this correspondent is, amongst other things, influenced by the weather situation. So, for example, if the sun is shining, he likes to get up early and therefore sends his report already before breakfast. If it is raining, on the other hand, he likes to sleep a bit longer and accordingly does not start his daily routine before he has had some cups of strong tea. Consequently, depending on the time the report arrives, the news agency can also make inferences about the weather situation in the foreign country. But, because this is not the only determining factor for the correspondent's working discipline, the time the reports arrive may only suggest a particular weather situation. So, in analogy to an HMM, the specific times of arrival of the reports can be seen as the sequences of the observable output symbols of the HMM generated by the emission probabilities of the hidden states – the underlying probable sequence of different weather situations.

---

[2] See also [14, p. 6f].

[3] See chapter 4, Type-3 Grammar, DFA or NFA.

Hidden Markov models can deliver continuous as well as discrete distributions of emission probabilities. In algorithmic composition, however, continuous models are of little importance as the observed emissions are in most cases note values with quantized parameters. A hidden Markov model represents a coupled stochastic process, due to the transition probabilities of the states in the Markov model and the state-dependent emission probabilities of the observed events. The following indications and symbols are used for the formal description of an HMM [12, p. 7ff]:

N	Number of states in a Markov model
$\{S_1,\ldots,S_N\}$	Set of these states
$\pi=\{\pi_l,\ldots,\pi_N\}$	Vector of *initial probabilities* for each state (*initial state distribution*)
$A=\{a_{ij}\}$	Transition probabilities in the MM from state to state
M	Number of observable output symbols of the HMM
$\{v_1,\ldots,v_M\}$	Set of output symbols
$B=\{b_{jm}\}$	Emission probabilities as probability of output of a symbol in a state
T	Length of the output sequence
$O=O_1\ldots O_T$	Sequence of output symbols with $O_t\in\{v_1,\ldots,v_M\}$
$Q=q_1\ldots q_T$	Sequence of state sequences in the MM at output O with $q_t\in\{1,\ldots T\}$

A hidden Markov process with the three states S_1, S_2, S_3 and the three output symbols V_1, V_2, V_3 can be graphically represented as shown in figure 3.3.

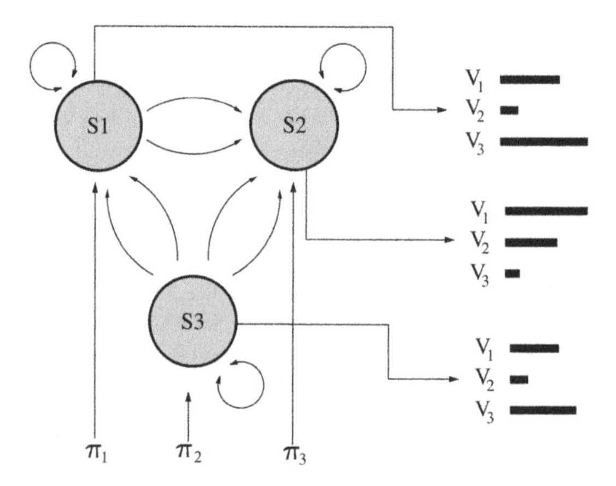

Fig. 3.3 Representation of a hidden Markov model.

In order to answer essential questions within a hidden Markov model, mostly three algorithms are applied:

- The *forward algorithm* computes the probabilities for the occurrence of a particular observable sequence, where the parameters (transition and observation probabilities as well as the initial state distribution) of the HMM are known.

- The *Viterbi algorithm* calculates the most likely sequence of hidden states, called the *Viterbi path*, on the basis of a given observable sequence.
- The *Baum–Welch algorithm* is applied to find the most likely parameters of an HMM on the basis of a given observable sequence.

3.3 Markov Models in Algorithmic Composition

In algorithmic composition, the transition probabilities of a Markov model are generated either according to individual structural parameters, or calculated in the process of generating style imitations by analyzing a corpus.

Application of Markov processes in musical structure generation was first examined by Harry F. Olson (1901–1982) around 1950. Olson was an American electrical engineer and physicist who focused on acoustic research. Together with Henry Belar, he developed the "Electronic Music Synthesizer" in 1955, the first machine to be called a synthesizer. Olson analyzed eleven melodies by Stephen Foster and produced Markov models of first and second order in regard to pitches and rhythm (indicated by Olson as dinote- and trinote probabilities) [13, p. 430 ff]. Figure 3.4 shows results of the statistical analysis in regard to the probabilities of rhythmic patterns in 4/4 and 3/4 time as well as the occurence of certain pitch classes. Figure 3.5 shows a transition table for pitch classes corresponding to a first order markov process. For the standardization of the analysis all songs were transposed to the key of D.

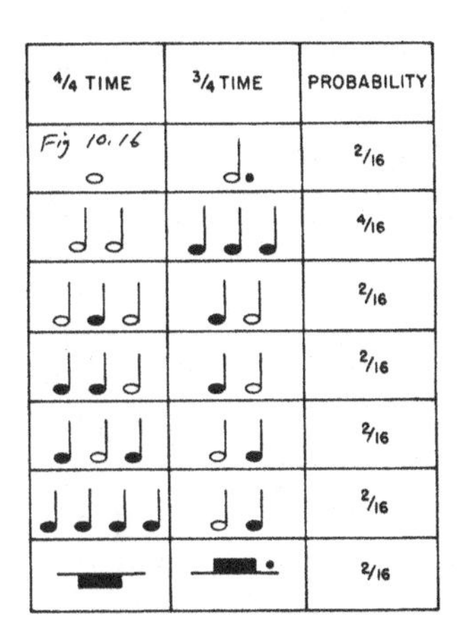

Note	B_3	$C^\#_4$	D_4	E_4	$F^\#_4$	G_4	$G^\#_4$	A_4	B_4	$C^\#_5$	D_5	E_5
Relative frequency	17	18	58	26	38	23	17	67	42	29	30	17

Fig. 3.4 Probability of rhythms (top) [13, p. 433] and relative frequency of the notes (bottom) [13, p. 431] in eleven Stephen Foster songs. Reproduced with kind permission by Dover Publications.

Probability of following note

Note	B_3	$C^\#_4$	D_4	E_4	$F^\#_4$	G_4	$G^\#_4$	A_4	B_4	$C^\#_5$	D_5	E_5
B_3			16									
$C^\#_4$			16									
D_4	1	1	2	5	3	1		1		1	1	
E_4		1	6	3	4			1			1	
$F^\#_4$			2	4	5	2		2	1			
G_4					4	3		6	3			
$G^\#_4$								16				
A_4			1		5	1	1	4	3		1	
B_4			1		1	1		9	2		2	
$C^\#_5$									8		8	
D_5								4	7	3	1	1
E_5								6		10		

Probability of note following the preceding note expressed in sixteenths.

Fig. 3.5 Two-note sequences of eleven Stephen Foster songs [13, p. 431]. Reproduced with kind permission by Dover Publications.

The difference between these two models can be seen in the fact that the productions of 2nd order Markov models show more harmonious melody creations as well as better results for the end part of the composition.

From 1955 on, Lejaren Hiller and Leonard Isaacson worked with the ILLIAC computer at the University of Illinois on a composition for string quartet: The "Illiac Suite" was performed for the first time in August 1956 and became famous as the first computer-generated composition.[4] Each of the movements, so-called "experiments," was dedicated to the realization of a special musical concept. In "experiment four," Hiller and Isaacson use Markov models of variable order for the generation of musical structure. Amongst others, these Markov models serve to select notes under various musical aspects, like the succession of skips and stepwise motions, the progression from consonant to dissonant intervals or even sound textures, which can be related to a tonal center in order to establish a distinct tonality.

Iannis Xenakis began to use Markov models for the generation of musical material[5] in 1958. In "Analogique A," Markov models are employed to arrange segments of differing density. Each of these segments, called "screens," consists of sounds of different dynamics. Figure 3.6 (top) shows four "screens" in which the lines represent a dynamic level and the columns a group of instruments. The transition probabilities of the "screens" result from a probability matrix for a 1st order Markov process as shown in figure 3.6 (bottom).

F.P. Brooks, A.L. Hopkins, P.G. Neumann and W.V. Wright [4] used a corpus of thirty-seven chorale melodies of similar metric-rhythmic structure for the Markov analysis and the subsequent generation. All chorales that are used for the generation model are in 4/4 time, begin on the last beat of the four beat measure, and do not contain note values shorter than an eighth note, which serves also as the basic rhythmic unit for the representation. Pitches within a range of four octaves are denoted by integers between 2 and 99, where all even numbers stand for a new note

[4] Cf. [8, p. 12], for the "Illiac Suite," also see chapters 2 and 10.

[5] Here: "Analogique A," "Analogique B" (1958) and "Syrmos" (1959).

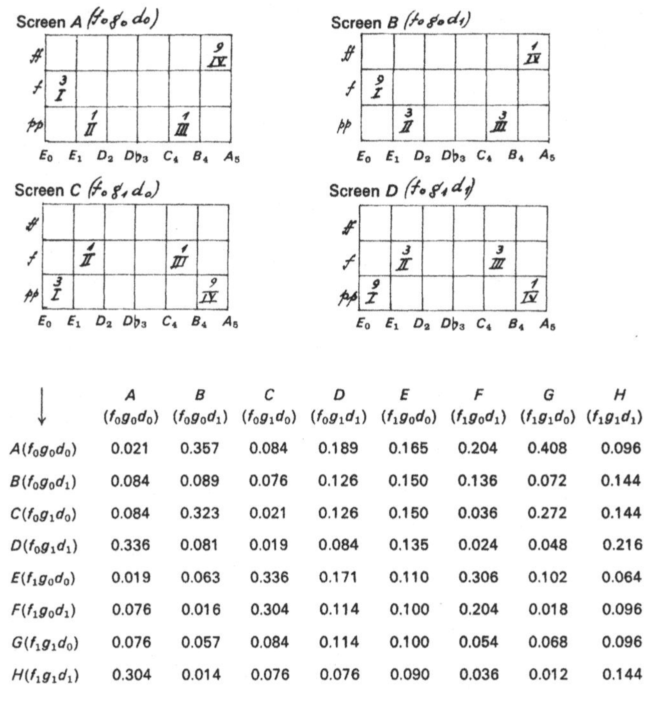

↓	A $(f_0g_0d_0)$	B $(f_0g_0d_1)$	C $(f_0g_1d_0)$	D $(f_0g_1d_1)$	E $(f_1g_0d_0)$	F $(f_1g_0d_1)$	G $(f_1g_1d_0)$	H $(f_1g_1d_1)$
$A(f_0g_0d_0)$	0.021	0.357	0.084	0.189	0.165	0.204	0.408	0.096
$B(f_0g_0d_1)$	0.084	0.089	0.076	0.126	0.150	0.136	0.072	0.144
$C(f_0g_1d_0)$	0.084	0.323	0.021	0.126	0.150	0.036	0.272	0.144
$D(f_0g_1d_1)$	0.336	0.081	0.019	0.084	0.135	0.024	0.048	0.216
$E(f_1g_0d_0)$	0.019	0.063	0.336	0.171	0.110	0.306	0.102	0.064
$F(f_1g_0d_1)$	0.076	0.016	0.304	0.114	0.100	0.204	0.018	0.096
$G(f_1g_1d_0)$	0.076	0.057	0.084	0.114	0.100	0.054	0.068	0.096
$H(f_1g_1d_1)$	0.304	0.014	0.076	0.076	0.090	0.036	0.012	0.144

Fig. 3.6 Screens A-D and probability matrix in "Analogique A" [16, p. 89, 101]. Reproduced with kind permission by Pendragon Press.

event and the uneven ones represent a note held over from the previous one. For the analysis and further generation, the different chorale melodies are transposed to C major. Markov analyses up to the 8th order are generated for the whole corpus, as shown in figure 3.7a, for example, the representations called "octograms" can be seen, for each eighth element with seven same predecessors (character strings of length n are generally referred to as "n-grams"); the following columns contain amongst others the relative frequency of each eighth element. According to the mapping, e.g. the number sequence of the first group (* 36 37 26 27 32 33 26 22) represents the tone pitches $G\overline{G}C\overline{C}A\overline{A}CD$, the notes under the bar indicating a tie. With a basis of sequences of 64 segments with a duration of each 1/8 per chorale melody, the result is a total of 2368 constellations for 37 chorales, of which only 1701 are different. In figure 3.7b, top, the number of the different n-grams for the starting notes $G\overline{G}F\overline{F}E\overline{E}D\overline{D}C\overline{C}$ are shown (in the figure, Brooks et al. use m instead of n). In figure 3.7b, bottom, all different n-grams with struck notes, held notes or rests as starting values are listed.[6]

In the generation of musical material, some other conditions are also established, e.g. on particular metric stresses no pauses are allowed, or melodies have to end at particular scale degrees [4, p. 33ff]. Figure 3.8 shows some examples of musical

[6] Brooks et al. do not explain the meaning of the hyphen.

Cell 1 2 3 4 5 6 7 8	Octo-gram Count	Hepta-gram Count	Relative Fre-quency	Cumula-tive Probability
36 37 26 27 32 33 26 22	1			
*36 37 26 27 32 33 26 22	2		2/8	2/8
36 37 26 27 32 33 26 27	1			
36 37 26 27 32 33 26 27	2			
36 37 26 27 32 33 26 27	3			
36 37 26 27 32 33 26 27	4			
*36 37 26 27 32 33 26 27	5		5/8	7/8
*36 37 26 27 32 33 26 28	1	8	1/8	8/8
36 37 26 27 32 33 32 33	1			
*36 37 26 27 32 33 32 33	2	2	2/2	2/2
*36 37 32 33 32 33 32 33	1	1	1/1	1/1
36 37 32 33 32 33 36 37	1			
*36 37 32 33 32 33 36 37	2		2/3	2/3
*36 37 32 33 32 33 36 42	1	3	1/3	3/3

a

Initial Note	Order of Analysis m							
	1	2	3	4	5	6	7	8
12 G	1	3	15	25	57 -	66	99 -	102
13 $\overline{G}$	1	12	30	67 -	84	124 -	131	150
16 F	1	6	14	25	43 -	45	63 -	63
17 $\overline{F}$	1	8	14	32 -	33	51 -	51	66
18 E	1	6	19	32	74 -	83	131 -	131
19 $\overline{E}$	1	11	23	68 -	78	130 -	135	156
22 D	1	4	19	30	66 -	69	104 -	109
23 $\overline{D}$	1	12	21	57 -	64	102 -	106	126
26 C	1	6	18	28	65 -	71	111 -	112
27 $\overline{C}$	1	13	25	65 -	72	112 -	113	136
All Struck Notes	18	47	152	219	444 -	479	698 -	705
All Held Notes	18	110	182	428 -	485	717 -	738	869
00 Initial Rest	1	5	10	28	45	70	95	127
Total Distinct m-Grams	37	162	344	675	974	1266	1531	1701

b

Fig. 3.7 Markov analyses and results in Brooks et al. [4, p. 32]. © 1993 Massachusetts Institute of Technology. By permission of The MIT Press.

production from 1st to 8th order. Furthermore, this study points out three problems that may occur in connection with structures generated by Markov models: Besides the danger of noticeable randomness of an output when applying a lower-order Markov model and overly restricted choices with higher orders, Brooks et al. also indicate a problem that may be easily overseen, which deals with the training corpus of the MM on whose basis the transition probabilities are determined. If the training data have a very similar structure, analyses of higher order may possibly be unnecessary, since, for instance, for each different transition probability of nth order only one possible transition of the order n + 1 may result. Therefore, an analysis of higher-order models does not contain any relevant information [4, p. 29].

Fig. 3.8 Production examples of different orders (m) by Brooks et al. [4, p. 38]. © 1993 Massachusetts Institute of Technology. By permission of The MIT Press.

3.3.1 Alternative Formalisms

An approach based on a production without previous analysis is described by Kevin Jones [11]. The author considers different relations that states of a MM may be in and establishes, based on the existent transition probabilities, "equivalence classes" of states within the Markov model. According to Jones, one class forms an accumulation of elements that are networked to a large extent, meaning that in their representation in a graph they can reach each other over the edges. Transitions between different classes are limited to only a few edges. Moreover, Jones distinguishes between "transient" and "recurrent classes" that are characterized by the fact that through the transition probabilities, states in "transient classes" may lead to states in other classes including "recurrent classes." If such a "recurrent class" is reached from "outside," it cannot be left any more, because the states within this class can only lead to states within the same class. Principally, this classification is a representation of an incomplete Markov model[7] where transition probabilities $p = 0$ are not taken into consideration and states which are related by transition possibilities $p > 0$ are interconnected by edges within a graph representation (figure 3.9).

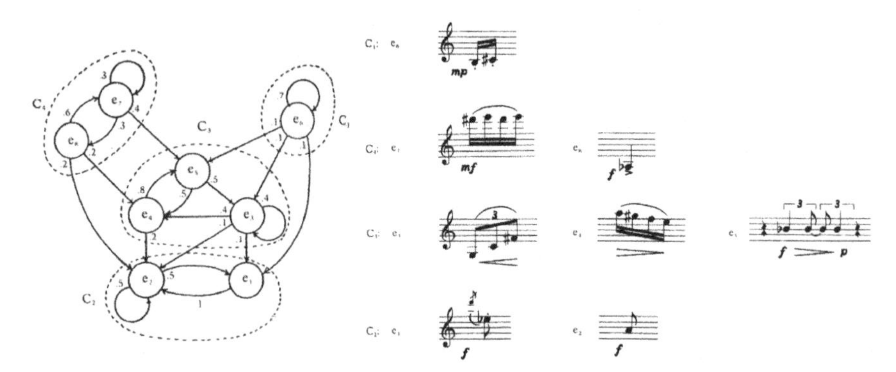

Fig. 3.9 Markov models in different equivalence classes (left), and terminals (right) [11, p. 384-385]. © 1993 by the Massachusetts Institute of Technology.

Jones's approach shows that for the states of a Markov model in the context of algorithmic composition, motivic structures instead of notes can also be used. Figure 3.10 demonstrates an example production of this model.

Fig. 3.10 Example production by the system of Jones.

In the comprehensive approach by Dan Ponsford, Geraint Wiggins and Chris Mellish [14], harmonic structure is generated by using a corpus of 17th-century French dance music. Here, Markov models are used to produce chord progressions. In choosing the corpus, the formal harmonic modeling of different dance forms was examined, among them music by Louis Couperin, Jean Baptiste Lully and Marin Marais. The authors finally decided to use the sarabande and selected 84 examples of this dance form for further examination. The musical material is divided by composer and mode into sub-corpora, while the harmonic progressions are represented

[7] See also the differently structured Markov models described by Chai and Vercoe in [5, p. 3].

as scale degrees. For further processing, the musical material is simplified in different ways, such as by rounding the note lengths or reducing the harmonic structure to triads in root position, where doublings and resolving dissonances are omitted.[8]

Responding to the problems occurring in higher-order Markov models, Ponsford et al. use "smoothed n-grams." A segmentation [14, p. 18ff] of the musical material into phrases and bars conducted for the training of the MM gives better results in the production. The beginning and the end of the particular pieces as well as smaller form sequences are identified in the training corpus so as to allow consideration of context dependencies for the generation. For the production, the transition probabilities are used according to the musical material, depending on the structure: "Numbering phrases effectively makes the data quite a lot sparser, as n-grams at phrase boundaries effectively operate over different sub-corpora, depending on the lengths of the piece they are from." [14, p. 20]. Therefore, for example, it is possible to distinguish between identical phrases (and bars) in chorales of differing length: "However, phrase 6 in a six-phrase piece would not have the same meaning as phrase 6 in an eight-phrase piece." [14, p. 20]. A number of compositions are generated with 3rd and 4th order Markov models, the experiments giving better results in 4th order MM. For the further improvement of the generated material, the authors suggest a correction of the applied Markov models by the user.

3.4 Hidden Markov Models in Algorithmic Composition

Martin Hirzel and Daniela Soukup [9] generated jazz improvisations based on small patterns, which are processed by an HMM. These melodic patterns are entered by the user, form the repertoire of the system and can be transposed to fit provided harmonic progressions. The HMM is trained with "Forest Flower (Sunrise)" by Charles Lloyd and after an input of harmonic progressions serving as the observable sequence, the Viterbi algorithm generates an appropriate succession of melodic patterns – representing the hidden states – that form the compositional output of the system.

Mary Farbood and Bernd Schoner [7] applied hidden Markov models to generate various counterpoints in relation to a given *cantus firmus* (Latin for "fixed melody"). In their work, the authors refer to the sixteenth-century counterpoint and in this context to the so-called first species, one framework of voice leading rules established by the Danish composer and musicologist Knud Jeppesen.[9] On the basis of these voice leading rules, an HMM is created, whose observable sequence is provided by the cantus firmus, and the Viterbi algorithm calculates the hidden states a succession of note values, which form the new counterpoints. An interesting approach in this work is the integration of different rules in one "unifying" transition table: "Each rule is implemented as a probability table where illegal transitions are described by

[8] For an explanation of the different reductions, see [14, p. 14ff].

[9] For the various rules, see Jeppesen's well-known book "Counterpoint: The Polyphonic Vocal Style of the Sixteenth Century" [10].

probability zero. The transition probabilities for generating a counterpoint line are obtained by multiplying the individual values from each table, assuming the rules are independent." [7, p. 3]. Figure 3.11 shows an example of the generation of two independent counterpoints according to a given cantus firmus (bottom).

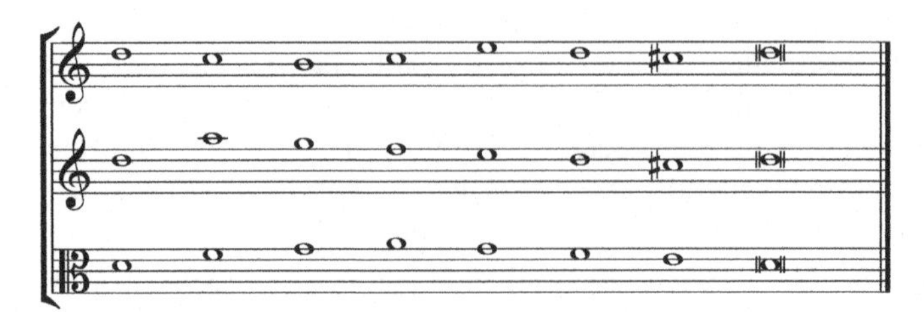

Fig. 3.11 Example generation by Farbood and Schoner according to a given cantus firmus.

3.4.1 A Hierarchical Model

Moray Allan [1] used Markov models and hidden Markov models for the harmonization of given soprano voices. Bach chorales are divided into training and test sets. The musical material is represented by means of a hierarchical segmentation[10] through the tone pitches of all voices, distinguishing amongst others between phrases and bars, the harmonic function also being annotated. The transition probabilities and emission probabilities are determined by frequency analysis of the reference corpus. In comparison to the HMM, Allan also illustrates the possibilities of conventional Markov models up to the 8th order that generate the harmonic function alternatively from the preceding harmonies, the melody tones or a combination of both. In the HMM considered best for the generation, the soprano voice forms the observable sequence in the HMM, while the underlying harmony corresponds to the hidden states. The Viterbi algorithm estimates the underlying state sequence and may also produce a universally acceptable structure in the range of a whole chorale. In contrast, here, conventional Markov models mostly fail to perform this task, because it cannot be estimated how the choice of a particular state transition at a particular time influences the whole structure.[11]

The harmonization is divided into three subtasks[12] that are solved by applying different HMMs. First, a harmonic skeleton is built, in which the concrete pitches and possible doublings are not indicated. The harmonic skeleton represents the hidden states, whereas the notes of the given soprano voice denote the observable sequence. Secondly, on the basis of the harmonic symbols as the observable sequence,

[10] For the representation of the musical material, cf. the scheme in [1, p. 18].

[11] See also Nearest Neighbor Heuristic in chapter 10.

[12] Cf. "Final harmonization model" in [1, p. 43ff].

another HMM generates concrete chords that correspond to the hidden states. Finally, a third HMM model is responsible for the ornamentation, whereas compound symbols, indicating the harmonic symbol and the notes of the current and next beat, denote the observable sequence. Each quaver associated to a harmonic symbol is represented as four distinct 1/16 notes, which form the hidden states and can – if partly altered by the HMM – result in various ornamentations. Figure 3.12 shows different examples of this three-stage generation process.

Fig. 3.12 Part of a three-stage harmonization process by Moray Allan [1, p. 49ff] according to the first chorale lines of "Dank sei Gott in der Höhe," BWV 288. With kind permission of Moray Allan.

3.4.2 Stylistic Classification

Wei Chai and Barry Vercoe [5] applied hidden Markov models as a means to distinguish the stylistic differences in music styles of different provenience. Although this approach is not used for the generation of musical structure, it contains interesting aspects of representation, worth to be mentioned in the context of this chapter. Chai and Vercoe represent the musical material in four different ways: First, absolute pitch representation within one octave space, starting with c1; secondly, absolute pitch representation with duration, where information on various durations is provided by eventually repeating each note multiple times; thirdly, interval representation, in which the pitches are converted into a sequence of intervals; fourthly, contour representation – a symbolic representation of the melodic contour, where the interval changes are denoted with the following symbols: 0 for no change, $+/-$ for ascending/descending 1 or 2 semitones, $++/--$ for ascending/ descending 3 or more semitones. The authors work with 16 different HMM, which are constructed out of different numbers of hidden states (2 to 6) in combination with four basic structures (see figure 3.13): "(a) A strict left-right model, each state can transfer to itself and the next one state. (b) A left-right model, each state can transfer to itself and any state right to it. (c) Additional to (b), the last state can transfer to the first state. (d) A fully connected model." [5, p. 3].

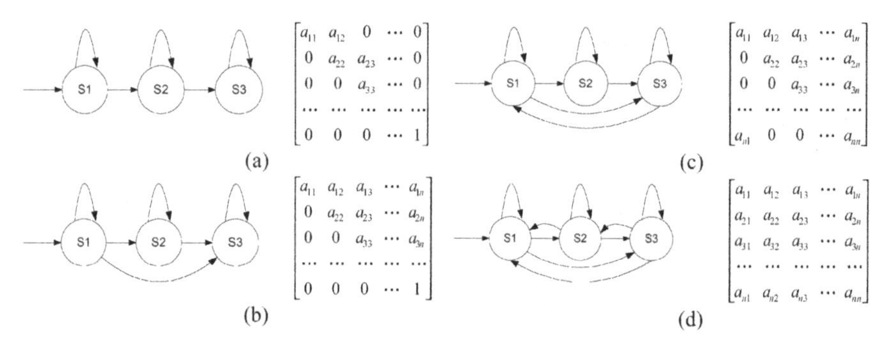

Fig. 3.13 Different types of Markov models by Chai et al. [5, p. 3]. With kind permission of Wei Chai.

For the analysis, the authors decide to use a corpus of monophonic folk music melodies from Ireland, Germany and Austria. The different folk melodies are separated into a training set (70%) and a test set (30%). The parameters of the HMMs are trained in regard to each country with the Baum–Welch algorithm, whereas the Viterbi algorithm is used to assign a melody from the test set to a particular HMM, thus allowing for a stylistic classification. The results point out structural similarities between German and Austrian folk songs and further demonstrate that stylistic classification may best be performed by HMMs with structure (a) or (b), whereas the number of states is of less importance.

3.5 Synopsis

Markov models are, like generative grammars, *substitution systems*, and due to their structure only allow the description of context dependencies through the transition probabilities of symbols in direct succession. These formalisms were originally developed in the context of natural language processing and are best suited to model one-dimensional symbol sequences. This structural feature of MMs does not correspond to musical information, which mostly adds a vertical dimension through layers, i.e. interconnected voices. Of course, a model defining all possible vertical constellations as single states could overcome this restriction, but if the required number of states is considered, this approach turns out to be a rather theoretical solution. A possible way to frame dependencies within a vertical structure consists in the application of hidden Markov models as coupled stochastic processes. Within the HMM, the Viterbi algorithm provides a suitable tool for the generation of an overall favorable structure, a demand, which can hardly be accomplished by the conventional formalism, except by approaches using several Markov models in a hierarchical interconnected structure. Since especially higher orders show a large size of the transition tables, these are mostly generated by analyzing an underlying corpus, leading to the fact that Markov models are in the majority of the cases used in the field of style imitation, with some restrictions: The danger of an obvious randomness with lower-order Markov models and also the frequent occurrence of an over-generation by models of higher order, meaning that large sections of the corpus are simply "re-generated." If the consistency of the corpus restricts the establishment of a higher-order Markov model, smoothed n-grams can be applied that use lower-order transitions for the generation of higher-order transition probabilities. An often overlooked deficiency in higher-order Markov models lies in their inability to indicate information which is provided in lower-order models. So, for example [15, p. 196–197] the symbol string AFGBBFGCFGDFG#EFG can be exhaustively described and therefore regenerated with a 3rd order Markov model, because there are no equal groups composed of three successive symbols in the corpus. This 3rd order Markov model provides an accurate description, which can actually be used for a regeneration of the corpus, but the apparent fact that an F is always followed by a G cannot – in contrast to a 1st order Markov model – be depicted.

Despite these disadvantages, Markov models are well suited to certain musical tasks. The type and quality of the output will depend largely on the properties of the corpus and can also be predicted very well in comparison to other procedures such as neural networks or cellular automata. Besides for style imitation, composers like Xenakis use Markov models in an innovative way, demonstrating their applicability also for the field of computer assisted composition, where this formalism is in general employed only rarely.

References

1. Allan M (2002) Harmonising chorales in the style of Johann Sebastian Bach. Thesis, University of Edinburgh, 2002
2. Basharin GP, Langville AN, Naumov VA (2004) The life and work of A. A. Markov. http://www.maths.tcd.ie/ nhb/talks/Markov.pdf Cited 24 Jul 2005
3. Bronstein IN, Semendjajew KA, Musiol G, Mühlig H (2000) Taschenbuch der Mathematik, 5th edn. Verlag Harri Deutsch, Thun und Frankfurt am Main. ISBN 3-8171-2015-X
4. Brooks FP, Hopkins AL, Neumann PG, Wright WV (1993) An experiment in musical composition. In: Schwanauer SM, Levitt DA (eds) Machine models of music. MIT Press, Cambridge, Mass. ISBN 0-262-19319-1
5. Chai W, Vercoe B (2001) Folk music classification using hidden Markov models. In: Proceedings International Conference on Artificial Intelligence, 2001
6. Dodge C, Jerse TA (1997) Computer music: synthesis, composition, and performance, 2nd edn. Schirmer Books, New York. ISBN 0-02-864682-7
7. Farbood M, Schoner B (2001) Analysis and synthesis of Palestrina-style counterpoint using Markov chains. In: Proceedings of International Computer Music Conference. International Computer Music Association, San Francisco
8. Hiller L, Isaacson L (1993) Musical composition with a high-speed digital computer. In: Schwanauer SM, Levitt DA (eds) Machine models of music. MIT Press, Cambridge, Mass. ISBN 0-262-19319-1
9. Hirzel M, Soukup D (2000) Project Writeup for CSCI 5832 natural language processing. University of Colorado, 2000
10. Jeppesen K (1992) Counterpoint: the polyphonic vocal style of the sixteenth century. Dover Publications, New York. ISBN 04862736X
11. Jones K (1981) Compositional applications of stochastic processes. Computer Music Journal 5/2, 1981
12. Knab B (2000) Erweiterungen von Hidden-Markov-Modellen zur Analyse ökonomischer Zeitreihen. Dissertation, Mathematisch-Naturwissenschaftliche Universität Köln
13. Olson H (1967) Music, physics and engineering, 2nd edn. Dover Publications, New York. ISBN 0-486-21769-8
14. Ponsford D, Wiggins G, Mellish C (1999) Statistical learning of harmonic movement. Journal of New Music Research, 1999
15. Todd PM, Loy DG (eds) (1991) Music and connectionism. MIT Press, Cambridge, Mass. ISBN 0-262-20081-3
16. Xenakis I (1992) Formalized music. Thought and mathematics in music. Pendragon, Stuyvesant, NY. ISBN 0-945193-24-6

Chapter 4
Generative Grammars

Generative grammars are powerful methods for algorithmic composition and musical analysis. The basic linguistic model [16] developed by Noam Chomsky in 1957, is the initial point for the application of this and other, more extended generative principles in musical tasks. The works of Roads, Steedman, Sundberg, Lerdahl, Jackendoff, and others from the 1970s on have created a strong interest in the application of generative grammars in the production and analysis of musical structure. Fields that often use generative grammars are traditional European art music, jazz, as well as music ethnology. Related formalisms, such as *augmented transition networks*, are used for example in David Cope's[1] approaches for automatically generating compositions conforming with a given musical style.

Fig. 4.1 Noam Chomsky. © Donna Coveney/MIT.

The formalisms that make up the basis of the abovementioned works refer mostly to the model of the Chomsky hierarchy. The basic principle for generating construc-

[1] See: "Experiments in Musical Intelligence" in chapter 5.

tions of a particular context by using rewriting rules was already applied before Chomsky. Mathematicians such as Axel Thue (1863–1914) or Emil Post (1897–1957) invented ground-breaking formalisms on the basis of whose principles John Backus and Peter Naur developed the *Backus–Naur form* in 1959 which, as a meta-syntax for context-free grammars led to the development of Algol 60 (short for algorithmic language), one of the first imperative programming languages.

The rewriting formalism of a generative grammar finds its parallels in different types of automata, Lindenmayer systems and Markov models. A specialized form of generative grammar in the field of linguistics is the *categorical grammar* or *C-Grammar*, a predecessor of the PS-Grammar, created by the Polish logicians Stanislaw Lesniewski in 1929 and Kasimierz Adjukiewicz in 1935 [21, p. 145ff]. In algorithmic composition this type of grammar is not very common, but it is used, for example, by Mark Steedman [42] as a tool for analyzing jazz chord sequences.

4.1 Generative Grammars as a Model of the Theory of Syntax

The theory of syntax as a sub-area of linguistics deals with the formal structure of compound sentences. This syntax aims to represent the principles and structures of possible sentence formations in a language. In this context, "language" can be a "natural" or "artificial" language, with sentences, or more generally, expressions consisting of symbol strings, whose structure follows certain rules. A particular expression in the given language is made up of a combination of units; the syntax of the given language allows for checking whether the given expression conforms to its rules, i.e., whether it is syntactically correct. When the same syntactic rules are employed for generative purposes, new and formally correct linguistic constructions come into being.

An essential criterion regarding sequences within a formal language is their *well-formedness*, signifying their correctness in terms of the syntactic rules – this, however, does not automatically imply that these "sentences" are semantically accurate, i.e. meaningful. Accordingly, for example, by defining a language as a sequence of words, a simple possibility to produce new expressions is their arbitrary concatenation. The basic condition here is that these words are part of the language; they form the *alphabet* and are subsumed in a finite *lexicon*. All possible expressions gained by combination make up the *free monoid* or *Kleene closure*. Depending on whether the alphabet contains an empty word chain, further distinctions are made. So, in the case of simple concatenation, for an underlying alphabet of e.g. (x,y,z), the result is symbol strings such as *xxy*, *xxyy*, *yy*, *zyx*.

If the alphabet is subdivided hierarchically, rules can be formulated that may produce context-dependent syntactic structures. One such early analytical approach is shown in the *immediate constituent analysis* of the American structuralist Leonard Bloomfield, published in 1933 in his book "Language": "The principle of immediate constituents will lead us, for example, to class a form like *gentlemanly* not as a compound word, but as a derived secondary word, since the immediate constituents are the bound form *ly* and the underlying form *gentleman*."[2] Searching for such immediate constituents as grammatical units, which create complex constructions, has been a common analysis procedure in American linguistics from the 1930s on and was above all used in the analysis of different Native American languages. In linguistics, a constituent is referred to as a word or a group of words functioning as a single unit within a hierarchical structure. By means of the immediate constituent analysis, sentences are divided into constituents by continuous segmentation, until each constituent consists of only a single word or a single meaningful part of a word. According to this theory, the constituent classes that arise from the analysis provide information on the structural contexts in the language under examination. Naturally, through this segmentation, pieces of information of the same semantic content but different word order cannot be recognized as equivalent expressions. This problem may also occur with other generative approaches. Chomsky counters this problem by means of grammatical transformations as well as making the distinction between *deep structure* and *surface structure* of a sentence [17] (see Grammatical Transformation).

If, however, in contrast to the analytical approach of the constituent analysis a recursive rule system is given for the definition of a language that is capable of producing well-formed expressions in a chosen context, one speaks of a *generative grammar*; regarding its basic formalism, a generative grammar is by no means confined to linguistic contexts: "Syntax is the study of the principles and processes by which sentences are constructed in particular languages. Syntactic investigation of a given language has as its goal the construction of a grammar that can be viewed as a device of some sort for producing the sentences of the language under analysis." [16, p. 11].

The generation of new sequences in a generative grammar occurs by means of *rewriting rules*, where symbols on the left-hand side of an expression are rewritten by symbols on the right-hand side. One must distinguish between symbols that can be further rewritten (*non-terminal symbols*) and symbols that cannot (*terminal symbols*). According to a linguistic context, lexical categories such as nominal phrases or verbal phrases as well as their units such as nouns, verbs, etc. are used as non-terminal symbols, whereas concrete words of a particular category form the set of terminal symbols.

A special form of such rewriting formalisms in the linguistic context is the *phrase structure grammar*, where single non-terminal symbols on the left-hand side are replaced by one or more non-terminal symbols of the right-hand side of the production rules. By *lexical insertion rules*, the resulting non-terminal symbols are replaced by the appropriate terminal symbols from the lexicon of the language, e.g. the concrete words of the resulting sentence(s), a process also known as parsing or derivation of the grammar.

So, the overall process begins with a (non-terminal) starting symbol on the right-hand side, continues with rewritings of all non-terminal symbols, and ends when the output symbol string consists entirely of terminal symbols. In the following example

[2] [14, p. 209] cited by [21, p. 169].

[19, p. 26–27], the abbreviations in the order of their occurrence are as follows: S (sentence), NP (nominal phrase), VP (verbal phrase), V (verb), PP (prepositional phrase), AP (adjective phrase), Adv (adverb), A (adjective), P (preposition), DET (article; for historic reasons, in generative grammar it is called the *determiner*), N (noun) .

S	→	NP	VP		
VP	→	V	(NP)	(PP)	
AP	→	(Adv)	A	(PP)	
PP	→	P	NP		
NP	→	(DET)	(AP)	N	(PP)

Now, as an example, the following terminal symbols are indicated for the particular categories:

N	→	man	girl	John
DET	→	a	the	
V	→	met	saw	
A	→	nice	good	quick
Adv	→	very	extremely	
P	→	in	for	to

In the tree diagram of this grammar, one of the possible *derivations* produces the following sentence (figure 4.2):

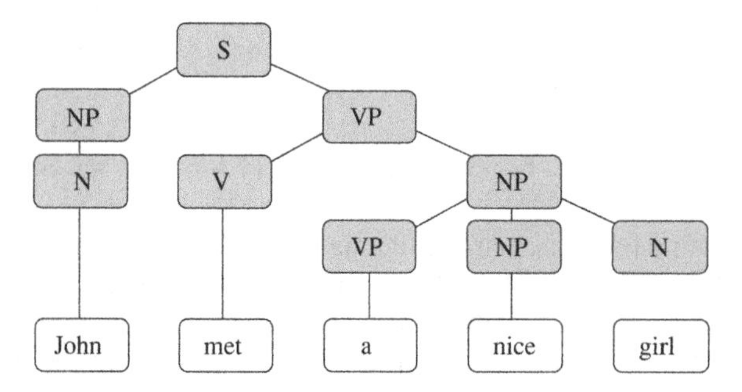

Fig. 4.2 Derivation of a PS-grammar.

One possible notation of a phrase structure grammar is a quadruple of the form (V, Vt, S, P), where the entries are identified as follows:

V	a finite set of non-terminal symbols
Vt	a finite set of terminal symbols
S	starting symbol of V without V_t
P	a set of rewriting and production rules of the form: $\alpha \rightarrow \beta$, where $\alpha \in V^+$ and $\beta \in V^*$
V^*	Kleene closure over V
V^+	positive closure or: V^* without the neutral element e (the empty word string)

4.1.1 The Chomsky Hierarchy

Chomsky distinguishes between four types of generative grammars that show different levels of restriction. These four types, starting with an unrestricted *type-0 grammar*, generate *formal languages* and correspond to different types of automata which may check whether a certain symbol string is part of the respective formal language and can thus be produced by the particular grammar. The higher a grammar's order is the more restrictive are the conditions for the application of its rules. The type of grammar is also related to the level of its *generative capacity* – so, a grammar's generative capacity is high, when it is able to generate several expressions and to prevent incorrect productions at the same time. However, a grammar's generative capacity is low if the rules only allow limited control over the expressions to be generated. Consequently, a grammar of lower order has a higher generative capacity, since in this case there are fewer limitations regarding the formulation of production rules. The generative capacity of a grammar is also directly related to its complexity, meaning the maximal effort required for the analysis of the expressions generated.

4.1.1.1 Type-0 Grammar (unrestricted grammar)

Restrictions: No restrictions; on both sides of the production rules, an arbitrary number of sequences of terminal or non-terminal symbols is possible.
Respective formal language: *Recursively enumerable language* or *partially decidable language*.
Respective automaton: *Non-deterministic Turing machine*. In a non-deterministic Turing machine, the same inputs can produce different possibilities for resulting state transitions.
Generative capacity: Very high. (Note that the indications "very high" to "low" only serve to compare generative capacities between the grammars in the Chomsky hierarchy.)
Complexity: Undecidable (up to infinite).

In investigating algorithmically whether an expression w is part of a language (L) generated by a particular grammar (G), this may be determined conclusively or not in a finite time or a finite number of computational steps. In this context *semi-decidability* means that for one combination of an input w, a language L and a grammar G, the membership of w can be determined beforehand, but in other cases the calculation may continue indefinitely without coming to a result. In the case of a Turing machine, this could mean that no final state (output) is reached and the calculation process continues infinitely. In this (worst) case, the complexity is therefore infinitely high.

4.1.1.2 Type-1 Grammar (context-sensitive grammar)

Restrictions: On both sides of the production rules, an arbitrary number of sequences of terminal or non-terminal symbols is possible, but the number of symbols on the right-hand side must not be smaller than the number on the left-hand side.
Formal language: *Context-sensitive language.*
Respective automaton: *Linear-bounded automaton.*
Generative capacity: High.
Complexity: exponential.

During the read/write operations, a linear-bounded automaton, as a restricted model of a Turing machine, never leaves the part of the tape bounded by the initial input. In general, context sensitivity refers to the possible comprehension of a context during rewriting, for example $AsT \rightarrow ArT$ meaning that the terminal symbol "s" is rewritten by the terminal "r" when embedded between the non-terminal symbols "A" and "T." All languages of type 1 and higher are *decidable*, i.e. for a given symbol string w it can be determined in a finite number of steps if this string belongs to $L(G)$. These languages are also referred to as *decidable languages*.

4.1.1.3 Type-2 grammar (context-free grammar)

Restrictions: The left-hand side of the production rules consists of one single non-terminal variable, the right-hand side of an arbitrary number of terminal or non-terminal symbols.
Formal language: *Context-free language.*
Respective automaton: *Pushdown automaton.*
Generative capacity: Middle.
Complexity: Polynomial.

A pushdown automaton is a finite automaton that can make use of a stack when choosing a transition path. Given an input symbol, current state and a stack symbol, this type of automaton can follow a transition to another state. Additionally, the stack may optionally be manipulated on top. In the beginning, the stack only consists of a single symbol indicating the end of the calculations. If, for example, a palindrome has to be recognized, the input symbols are written on the stack until one symbol is repeated. Then, when the input and current stack content are successively identical, the current symbols in the stack are deleted, until the symbol for the end of the calculation is reached. In this case, a palindrome can be successfully recognized by the pushdown automaton. By means of this function principle, context-dependent structure may be processed; this is not possible with more simple types of automata due to their lack of "knowledge" about preceding input symbols.

4.1.1.4 Type-3 grammar (regular grammar)

Restrictions: The left-hand side of the production rules consists of only one variable (non-terminal of V); on the right-hand side there is a terminal, followed by one non-terminal at most. This form of production rules is also referred to as *right-linear*. If there is a terminal on the right-hand side that is preceded by a non-terminal, these production rules are also called *left-linear*.
Formal language: Regular language.
Respective automaton: *Deterministic finite automaton* (*DFA*) or *non-deterministic finite automaton* (*NFA*).
Generative capacity: Low.
Complexity: Linear.

A DFA is a quintuple (S, Σ, δ, z_0, F), consisting of

S	a finite set of all states
Σ	a finite set called the input alphabet
z_0	a starting state
F	set of final states ($F \in S$)
δ	transition functions ($S \times S \rightarrow S$)

Given the following conditions [39, p. 27ff]

$S = \{z_0, z_1, z_3, z_3\}$				
$\Sigma = \{a, b\}$	$\delta(z_0, a) = z_1$	$\delta(z_0, b) = z_3$	$\delta(z_1, a) = z_2$	$\delta(z_1, b) = z_0$
$F = z_3$	$\delta(z_2, a) = z_3$	$\delta(z_2, b) = z_1$	$\delta(z_3, a) = z_0$	$\delta(z_3, b) = z_2$

this automaton can investigate in symbol strings if these are elements of the Kleene closure and therefore correct constituents of the respective language. In the recognition of single symbols, the automaton passes in the transition functions a sequence of different states z_0, $z_1, \ldots z_n$ (the symbols of Σ), z_0 meaning the starting state, until it reaches a final state $z_n \in F$. Therefore, the DFA will recognize an input of:

aaa as a correct symbol string of the respective language – corresponding to the resulting state transitions $\delta(z_0,a) = z_1$, $\delta(z_1,a) = z_2$, $\delta(z_2,a) = z_3$, in which the final state is also reached. As another example, the symbol string *bb*, however, leads from $z_0 \rightarrow z_3 \rightarrow z_0$, but because z_3 is valued as final state, *bb* will not be recognized as a valid member of the respective language.

A directed, marked graph (figure 4.3) represents the behavior of the DFA. The different states are represented by circles. The directed edges indicate the input symbols, for example: In z_0, the automaton transfers to z_1 at a given state *a*.

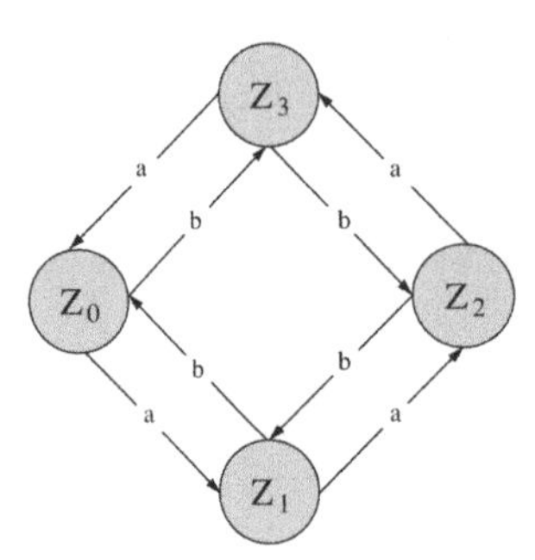

Fig. 4.3 State graph of a DFA.

Given the same input symbol, a non-deterministic automaton can transfer to different states. In the recognition of symbol sequences, the sequence is regarded as part of the language only if one of the possible final states is reached. For a correct definition of such an automaton, it is necessary that there exists at least one valid state sequence ending in a final state. Furthermore, a set of starting states is possible. In the notation as a quintuple, z_0 is substituted by the set of starting states. Type-3 grammars can also be represented by Markov models.

4.1.2 Grammatical Transformation

Chomsky meets the PS-grammatical representation of identical expressions of different word order by means of grammatical transformations that take phrase structure trees as an input and output them as transformed phrase structure trees. For the indication of the semantic content of a sentence that may be represented in different word orders, Chomsky uses the term "deep structure." Hence, the deep structure is a semantic basis for all expressions that can be produced by grammatical transformation and result in various mental representations of linguistic expressions referred to as "surface structures."

The formalism of a generative grammar can be employed on a large scale in applications of algorithmic composition. However, the concept of surface and deep structure and the semantic content of a syntactic expression cannot be applied to musical principles offhand. A grammatical transformation carried out for different word orders of a sentence of the same meaning will, applied to a musical structure, result in a completely new composition. The semantics of a linguistic expression of different transformation is reflected in the musical equivalent only by using the same terminals – a fact that does not allow a mutual criterion of the generated structures in the sense of "musical semantics" to be derived. A correspondent to linguistic surface and deep structure can be found most likely in the examination of musical relations in Heinrich Schenker's analysis methods and related approaches. Here, for example, a single underlying harmonic skeleton may serve as a basis for different forms of musical structures.

When musical analysis is mentioned in the following, it is based on the formalism of a generative grammar. If a generative model of the musical structure may be generated by analytical investigation of a corpus, it can also be used for further generations. In some approaches aiming at the generation of style imitations, the preceding analysis of style-compliant material is an essential precondition. In some other works mentioned, the generative aspect is not examined explicitly, but it is given implicitly by the grammatical model.

4.2 Generative Grammars in Algorithmic Composition

In applications of generative grammars for musical production and analysis, some prominent tendencies can be distinguished. Music ethnology aims at describing different genuine music styles by grammatical models; in European art music, hierarchical grammatical structures are often used in analysis and generation; in jazz, this formalism is frequently used to create chord progressions on the basis of musical corpora and rules of jazz harmony.

Concepts that allow a hierarchical division of the musical material and work with the substitution of symbols can naturally be best formulated by a generative grammar. If explicitly formulated rules are assumed, a *knowledge-based* approach is involved that is also applied in expert systems and comparable procedures. If a system automatically generates rewriting rules out of a corpus, it is a *non-knowledge-based* approach and can also be referred to as *grammatical inference*. Here, a comparison to further procedures may be drawn. For example, the expressiveness of Markov models equals to type-3 grammars. Due to the fact that they only allow for the treatment of a context of successive symbols, they are inferior to grammars of lower order. Another disadvantage of Markov models results from their fixed order, which is set before the model is generated and, in most cases, is not able to describe the context sufficiently. Artificial neural networks or genetic algorithms may also be used for the treatment of tasks where no domain-specific knowledge exists.

Systems that can independently find regularities in given data can also be found in the field of unsupervised learning of AI. These and similar algorithms can represent useful tools for the analysis and modeling of musical styles, on which there is insufficient domain-specific knowledge, but which may, however, be performed due to some kind of implicit rule system. Here, the generation of a terminal alphabet is necessary for the analysis of a corpus and the further generation of new musical material. In this case, terminals can be for example chords, harmonic movements, melodic fragments, rhythmic figures or also playing techniques of a particular instrument.

A musical structure generated by a grammar will fulfill by definition the criterion of being well-formed, but the structure does not necessarily have to comply with the implicit musical rules in every case. When a grammar produces rewriting rules on the basis of the state transitions of musical units in the corpus, this does not guarantee that this procedure does not break any implicit musical rules that can only be covered by treating a wider context. If, for example, a grammar has encoded a lot of transitions of musical fragments from the corpus in its rewriting rules, this does not necessarily mean that each of the generated compositions based on these, has to be correct according to the underlying style. A concrete example would be the obligatory use of a major triad independent of the predominant tonality – as final chord in a composition of a particular music style. Within the grammar, changing from minor to major could be formulated as a possible rewriting rule. An application of that rule would, however, only be correct at the end of the musical piece; at every other position, this nevertheless well-formed expression would break a musical rule which was not recognized by the grammar. If, to solve this problem, the formalism delivers a long-range description of the context within the corpus, there is danger of 'over-generation': The grammar simply regenerates large parts of the corpus, a tendency that can also be detected amongst others in higher-order Markov chains.

A second problem that can occur in the application of generative grammar without domain-specific knowledge may result from the existence of musical rules and basic conditions that cannot be covered by a grammatical inference at all. If the corpus consists of a number of correct examples that, however, implicitly represent different classes of musical material, this fact will not be able to be covered by the generated grammar. An example would be melodic movements possible within a particular key that are not felt to be style-compliant in another. A grammatical model that, for example, disregards keys through an intervallic representation of the different compositions of the corpus, could not consider these facts per se and would therefore produce a number of incorrect variations in the respective style. Besides these factors, extra-musical conditions that may have influence on the selection and composition of the musical material can also be considered. Particularly in the field of music ethnology, a great number of sociological, cause-conditional and ritual factors may greatly influence the musical performance and can hardly be encoded by a generative grammar, an aspect which is particularly emphasized in the work of John Blacking.[3]

If a grammar is produced on the basis of a corpus, an examination of the generated material is recommended. This evaluation is of particular importance for all non-knowledge-based systems of algorithmic composition, be that a human fitness

---

[3] See section "Music Ethnology."

rater, the different forms of feedback within supervised learning or an algorithmic evaluation of the output.

4.2.1 Musical Analysis by Generative Models

One important predecessor of generative grammar for application in music is Heinrich Schenker's (1868–1935) musical analysis methods [38]. According to Schenker, the components of a tonal structure may be referred to as an imaginary fundamental pattern he calls the "Ursatz," whose further structuring creates the different levels of a composition. Consecutively, Stephen Smoliar, James Meehan and Célestin Delige deal with an informatic treatment of Schenker's analysis methods. Inspired by Heinrich Schenker's ideas, in "A Generative Theory of Tonal Music," [27] Fred Lerdahl and Ray Jackendoff described an exhaustive model for the representation of tonal music by a generative formalism. An examination of musical representation by generative grammars including extended formalisms is also provided by Curtis Roads [36, 37].

According to Schenker's analysis, the "Ursatz" is an abstract two-voice structure, consisting of the "Urlinie" or fundamental descent, and the "Bassbrechung" or bass arpeggiation. The "Urlinie" is a linear movement within a harmonic progression, from a "Kopfnote" or "head-note" to the root of a target chord. The possible "head-notes" for a piece of music are third, fifth and octave of a triad. The "Bassbrechung" indicates the basic tones of the harmonic progression.

Fig. 4.4 "Ursatz" consisting of "Bassbrechung" (bottom) and "Urlinie" (top).

This "Ursatz," forming the musical "Hintergrund" ("background") of a piece of music is further structured by the "Ausfaltung" or "unfolding." The levels that result from this process are the "Mittelgrund" ("middleground") and finally the "Vordergrund" ("foreground") of the musical form, nearly corresponding to the actual notated score. The "Ausfaltung" is made by a process of "Auskomponierung" or "composing out" that occurs in a similar fashion to the different possibilities of setting voices according to a given cantus firmus.[4] The basic techniques are complete or incomplete arpeggiation, the connection in steps of tones belonging to a chord, as

[4] For a well-known work on counterpoint referring to the style of Giovanni Pierluigi Palestrina (ca. 1514–1594), see [23].

well as the insertion of upper and lower neighboring notes. These "diminutions" are also applied multiple times and finally produce the musical "foreground."

Stephen Smoliar [40] developed a comprehensive framework for automated analysis according to Schenker's method. This system enables, amongst other things, transformations of tree structures that represent different appearances of musical "deep structures" and facilitates the application of various techniques of unfolding of the musical material.

As Schenker's analysis method is controversial, some aspects of this approach are subject to a critical review, as shown in the work of Meehan [30] and Deliège [20]. The criticism concentrates especially on the construction of the "Ursatz" that according to Meehan is rather a theoretical construct of little musical significance. Deliège even reduces the meaning of the "Ursatz" to a solely symbolical value, similar to the starting symbol "S" of a generative grammar.

Lerdahl and Jackendoff made an analytic approach for examining music which can be described in a context of harmonic functions.[5] In comparison to Schenker, the hierarchical structure is divided more finely and it is of form-producing importance on all levels. For the purpose of an analysis, the musical material is treated under the following aspects: The "grouping structure" divides the piece into units such as motives, phrases and sections. The "metrical structure" controls the sequence of stressed and unstressed beats on different hierarchical levels. The "time-span reduction" assigns a structural importance to the pitches in the "grouping structure" and "metrical structure." The "prolongation reduction" ascribes the tone pitches a hierarchical importance in terms of their "tension," "relaxation" or "duration." "Well-formed rules" describe all well-formed expressions, while "preference rules" represent the structural selection criteria of an "experienced listener": "We have found that a generative music theory, unlike a generative linguistic theory, must not only assign structural descriptions to a piece, but must also differentiate them along a scale of coherence, weighting them as more or less "preferred" interpretations. [...] The preference rules, which do the major portion of the work, of developing analyses within our theory, have no counterpart in linguistic theory; their presence is a prominent difference between the forms of the two theories." [27, p. 9]. With the introduction of their preference model, Lerdahl and Jackendoff have created an interesting method for examining the musical generations of generative grammars beyond their aspect of well-formedness. As an example of a "time-span reduction," figure 4.5 shows a segmentation of the beginning of Beethoven's Sonata Op. 31, Nr. 2.

If structural generation is performed in the framework of an automated process on an existing musical corpus, the analysis and segmentation of the material must be made in advance. On these terms, David Temperley and Daniel Sleator developed an approach [45] for automated musical analysis, based on Lerdahl's and Jackendoff's ideas.

A procedure for automated segmentation of tonal music is described by Bryan Pardo and William Birmingham [33]. Their system "HarmAn" recognizes the un-

---

[5] For an introduction to this concept, see [28]; for a comparison of some of the mentioned approaches dealing with the "Urlinie," see [29].

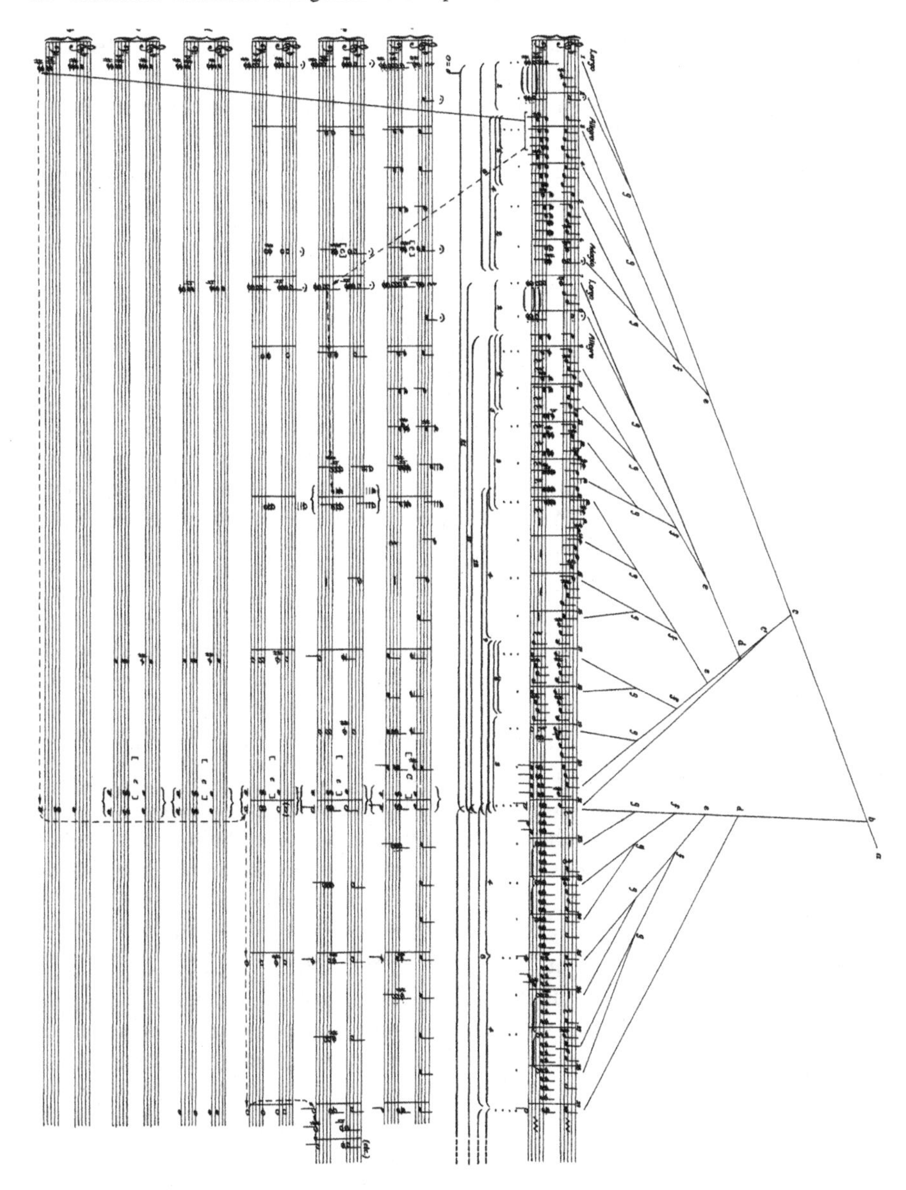

Fig. 4.5 "Time-span reduction" in Beethoven's Sonata Op. 31, Nr. 2 [27, p. 256]. © 1993 Massachusetts Institute of Technology. By permission of The MIT Press.

derlying harmonic function of a passage of tonal music by comparing patterns of preferred interval constellations.

Although music analysis is not in the scope of this book, the works that are treated as examples enable interesting possibilities in algorithmic composition, especially in the field of generative grammars, where a valid analysis model can also mostly be used for the generation of musical material. Other aspects in this regard can also be found in computationally based models of music cognition.[6]

4.2.2 Folk Music and European Art Music

An early approach joining harmonic and melodic aspects in the modeling of a musical genre is treated by Gary Rader [35]. He used generative grammars for the production of circle rounds, which are perpetual canons that return to their beginning and can be repeated infinitely. Additionally, this approach uses weighted probabilities in the application of the production rules. A weighting may, for example, prefer a step in a melodic progression, while a superordinated rule guarantees that within the movements no parallel fifths are produced. The system goes through a two-level process where first a harmonic structure is generated, out of which then the melodic material is developed.

The scale degrees I, II, III, IV, V and VI serve as a basis for the harmonic skeleton and succeed each other according to the probabilities A% to Z%, as can be seen in a graph in figure 4.6. The rules in the generation of the harmonic progressions constitute conditions such as a limited range of particular starting and end chords or harmonic functions on stressed beats. The harmonic skeleton is built up of chords in root position, whereas inversions are later applied for the construction of the melodic structure.

The generation of the melody occurs within a frame interval of two octaves and rhythmic values consist of eights notes or their multiples. A number of rules that can be paraphrased with general principles of counterpoint, structure the progression of durations and pitches. For the formal representation of the system, Rader uses a stochastic generative grammar which processes information on the already generated units and whose rules can be weighted. Figure 4.7 shows examples for generations of three-part rounds. The entry of the second and third voice is indicated with the beginning of a separate line.

Johan Sundberg and Björn Lindblom used generative grammars in the production of Swedish children's songs and, in another experiment, generated further variations of already existing versions of a folk song [43]. For their first experiment, eight-bar children's songs of the 19th century are analyzed in order to generate a grammar for the production of the songs. Figure 4.8 shows the results of the analysis in an eight-bar time line, where the distributions of musical elements are represented as hatched fields.

---

[6] See as an interesting example the comprehensive work of Sven Ahlbäck on the cognition of surface structures of monophonic melodies [1].

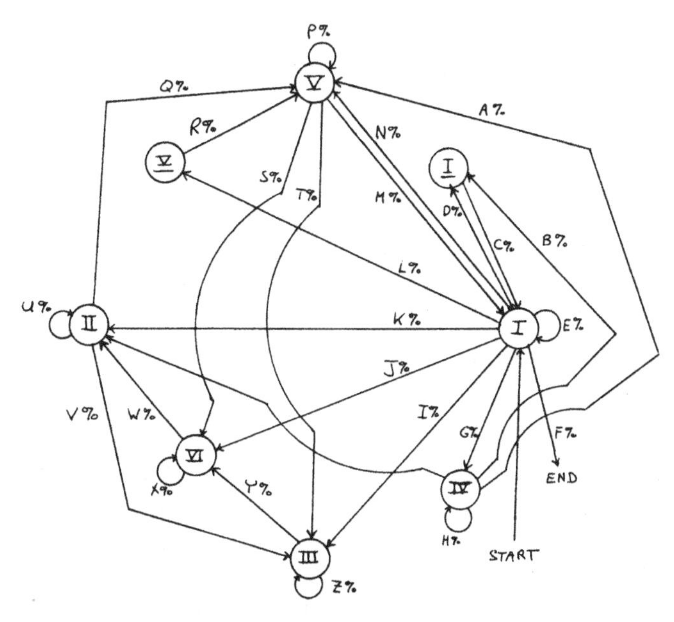

Fig. 4.6 Possible chord progressions for the generation of rounds [35, p. 249]. © 1993 Massachusetts Institute of Technology. By permission of The MIT Press.

Fig. 4.7 Example productions of three-part rounds by Rader.

According to Sundberg and Lindblom, "introductory chords" designate the tonics, "target chords" are chords that are preceded by their dominants, and "anticipatory chords" are dominants followed by the associated tonic. "Suspensions" are a special case of non-chord tones and always occur at temporal positions of short duration. A segmentation of the eight-bar structure into units hierarchically belonging together up to phrases, bars, and single notes, is brought into agreement with the results of the analysis. So, for the particular units, weightings referred to as "promi-

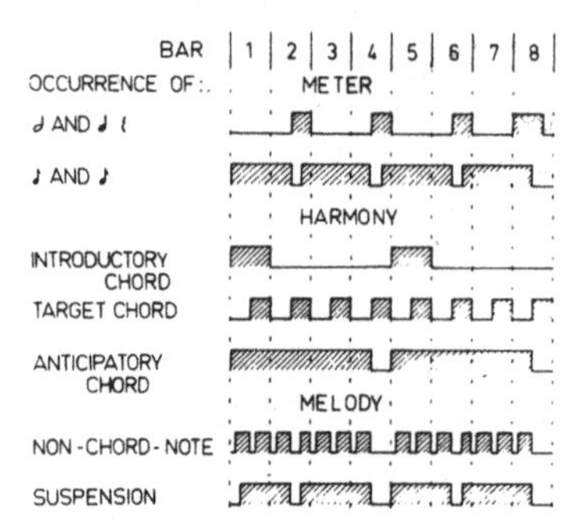

Fig. 4.8 Analysis of distribution of musical structure [43, p. 267]. © 1993 Massachusetts Institute of Technology. By permission of The MIT Press.

nence ranks" result which determine the application of different production rules. After the generation of concrete durations, the system creates harmonic structure on the bar level. The combination of rhythm and harmony finally allows, based on the analyzed data and further rules, the generation of the melodic line.

In another experiment, Sundberg and Lindblom used a similar procedure to generate further variations of a Swedish folk song that already existed in eight variations [43]:"[...] our rule system should generate only those melodies that are felt to be melodically similar to the versions given." [43, p. 279]. Although the "feeling for the similarity of variations" is a fuzzy quality criterion, the fact that the system produces variations that meet this task, proves the applicability of generative grammar in musical domains with little or no domain-specific knowledge. Figure 4.9 shows some original variations of the folk song, figure 4.10 four variations generated by Sundberg and Lindblom.[7]

In their different projects, Mario Baroni, Rosella Brunetti, Laura Callegari and Carlo Jacoboni [2] dealt with the generation of musical structure in different styles.

In the project MELOS 2, Lutheran chorale melodies are taken into consideration for the generation by a generative grammar, where two classes of rules are applied. Based on a frame interval, "micro formal rules" create melodic phrases in a multi-stage rewriting process. Then, "macro formal rules" are applied to put these phrases together to a chorale. What is noticeable in "MELOS 2" is its conceptual closeness to Schenker's concepts such as the "Urlinie" or the different techniques of "unfolding."

In their project CHANSON, similar principles and further possibilities are considered using a corpus of French dance music of the 18th century for the structuring of the melodies.

As an extension of MELOS 2, the project HARMONY generates a harmonization and a bass voice for a given chorale melody. The starting point for the generation of this grammatical model is the analysis of non-modulating segments of

[7] In [44], Sundberg and Lindblom compare their formalism to similar works such as [27], [2], [15] in the field of generative grammars; particular attention is paid to their hierarchical system of "prominence ranks."

Fig. 4.9 Some variations of a Swedish folk song.

Fig. 4.10 Variations additionally generated with generative grammar.

Lutheran chorales from J.S. Bach. The model is initialized with a starting chord and a final chord, determines positions for the tonics and further applies rewriting rules of possible chord substitutions, as shown in figure 4.11.

Finally, a bass line is produced that is designed according to similar principles as melody generation in MELOS 2.

An approach similar to MELOS 2 which is also characterized by starting and final chords as well as the further application of rewriting rules, is described by Lelio Camillieri [15]. The corpus consists of initial phrases in major keys from the Lieder cycles op. 25, op. 23/3 and op. 89 of Franz Schubert. According to Camillieri, generation, verification and correction of the grammar allow for the production of similar style-compliant phrases, but for modeling beyond this, this approach is not taken into consideration.

4.2.3 Music Ethnology

David W. Hughes [22] compared the generative capacity of a grammar developed by A. Becker et al. with one of his own approaches for the domain of Central Javanese

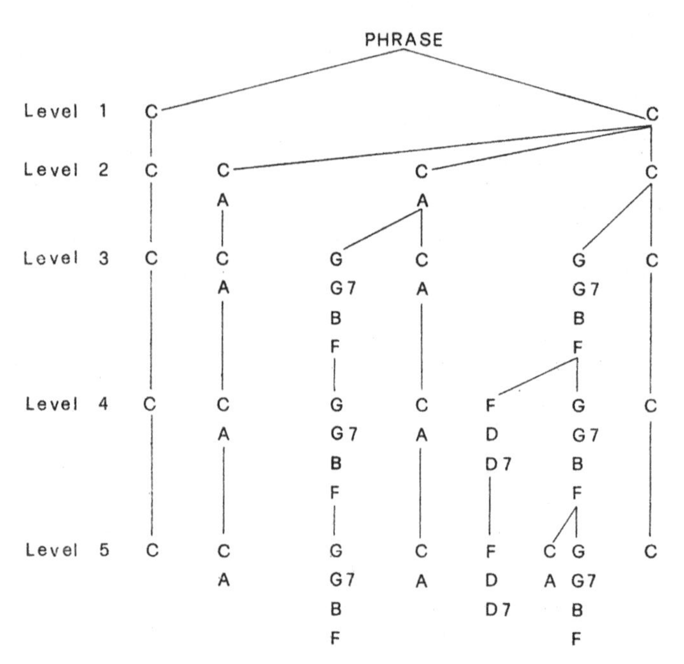

Fig. 4.11 Harmonic substitution in HARMONY [2, p. 212]. With kind permission of Casa Editrice Leo S. Olschki.

Gamelan music. As further examples, he gave an overview of different applications of generative grammars in music ethnology, comprising amongst others North Indian tabla music, rhythmic structures in Afghan lute music, South African vocal and instrumental music as well as a special type of Inuit song.

With Gamelan music, Judith and Alton Becker [8] created a grammar based on nine rules for a "Balungan," a "core melody" acting as a melodic skeleton that is further elaborated by the instruments of the Javanese Gamelan orchestra, in the context of a particular musical style. The problem with this grammar is that Becker comes from a corpus of only nine such core melodies meaning that this corpus is hardly suitable for representing the style sufficiently. However, in the variations of the Swedish folk songs mentioned before, this problem does not exist, since the musical range is limited by making variations of only one particular song.

David Hughes bases his grammar on a set of rules that he divides into five categories. "Base rules" control basic decision processes such as the choice of the style, the number of "gongans" or the choice of distinct scale tones. A "gongan" is a musical unit of a particular number of "gatras," end-weighted groups of four musical events that roughly correspond to bars in occidental music tradition. "Contour assignment rules" assign particular melody contours to the gatras. "Restriction rules" are limitations in the application of "transformation rules" that specify the melody contours by means of concrete note values. "Derivation rules" regulate form-specific

[8] [4], treated in [22, p. 333ff].

processes within each style. In Becker's and Hughes's works, the knowledge-based approach is of primary importance; in these cases, the generative grammar is more a formalism for the representation of the rules.

John Baily[9] developed a grammar for examining aspects of various plucking techniques of the Afghan lute. This approach also points to the works of Kippen and Bel, who created, on the basis of playing techniques of the North Indian tabla, grammars for the generation of pieces of a particular repertoire of this instrument (see next section).

In a number of works,[10] John Blacking dealt with vocal and instrumental melodies of the Venda, a genuine ethnic group in the South African Transvaal. Blacking develops rules such as for the creation of new metrical patterns or the preference of particular chords depending on their use in instrumental or vocal music, and also formulates characteristics of a musical deep structure, like relations of melodic and harmonic form. Blacking makes the interesting point that analysis and modeling should consider a number of aspects which do not directly shape musical structures, but are nonetheless essential for the exhaustive description of a distinct style. These aspects include, for example, the sociological and social context or conditions of the performance, which may have an influence on the realization of the musical style [13, p. 366ff].

Ramòn Pelinski [34] described a generative grammar for the "A ja jait" of the Inuit of the Hudson Bay. The "A ja jait" is a special song genre that recalls special moments in the life of a person and is in a spiritual sense strongly connected to the performer. Pelinski provides the following restrictions to his grammar: The grammar only treats the rhythmic and melodic structure of the genre. Different singing techniques as well as formal principles regarding volume and timbre are not taken into account. Furthermore, the context of the performance, as well as the relation between text and melody, is not treated in the grammatical analysis and generation. The song genre is divided hierarchically in formal units corresponding to introduction formulae, ending formulae and other parts of a similar formal significance. Rhythmic patterns, as well as allowed intervals and ornamentations, form the terminal alphabet of the grammar. The songs are divided into different classes, characterized by formal and modal aspects. A representation of the modes with the permitted melodic movements is shown with a graph in figure 4.12.

4.2.4 Bol Processor

In their investigations of North Indian tabla drum music in the 1980s, Bernard Bel and Jim Kippen developed an interesting approach to the application of generative grammars in the field of music ethnology. For the analysis of musical input and also for the generation of new improvisations, Bel designed the Bol Processor (BP)

---

[9] Cf. [3], treated in [22, p. 350ff].

[10] [12] and [13], treated in [22, p. 350ff].

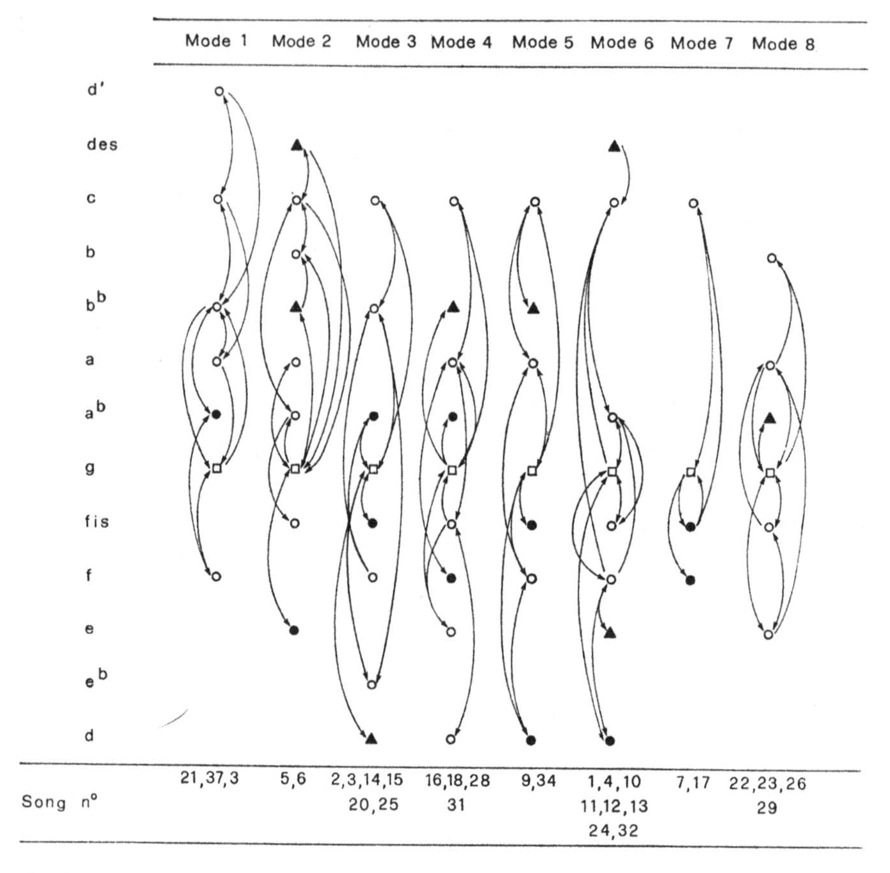

Fig. 4.12 Graph for interval classes of an Inuit song [34, p. 281]. With kind permission of Casa Editrice Leo S. Olschki.

computer system. Initially, this software consisted of a simple word processor for the notation of rhythmic patterns and was further developed over the years into an effective system of algorithmic composition based on generative grammars.[11]

Tabla music is bound to a number of rules and restrictions that are not explicitly formulated. The repertoire is not noted in writing but is represented in an oral notation system; however, rhythmic phrases may be denoted using onomatopoetic syllables. These syllables are called *bols*[12] (for example "dha," "thi," "trkt," etc.), each referring to a particular stroke or phrase on the instrument and consequently forming the basic repertoire of the traditional playing technique. The investigations of Bel and Kippen are based on a style of Indian tabla drumming known as *qa'ida*, which is a formal model that works with themes and variations. Table 4.1 shows

[11] For the works of Kippen and Bel as well as for the range of functions of the BP, see [5], [6], [7], [8], [9], [10], [11]; free software available at: http://bolprocessor.sourcefourge.net/.

[12] From the Urdu/Hindi "bolna" meaning "to speak"; cf. [7, p. 2].

dha tr kt dha	tr kt dha ge	dha ti dha ge	dhee na ge na
dha tr kt dha	tr kt dha dha	dha ti dha ge	dhee na ge na
dha ti dha tr	kt dha tr kt	dha ti dha ge	dhee na ge na
dha ti kt dha	ti-dha ti	dha ti dha ge	dhee na ge na
dha ti kt dha	ti dha tr kt	dha ti dha ge	dhee na ge na
ti-dha ti	dha dha tr kt	dha ti dha ge	dhee na ge na
ti dha tr kt	dha dha tr kt	dha ti dha ge	dhee na ge na
tr kt dha ti	dha dha tr kt	dha ti dha ge	dhee na ge na
tr kt tr kt	dha dha tr kt	dha ti dha ge	dhee na ge na
tr kt dha tr	dha dha tr kt	dha ti dha ge	dhee na ge na

Table 4.1 First lines of variations of a qa'ida.

the first ten lines of variations of a qa'ida that is always read line-by-line from left to right. A variation consists of sixteen bols of the same length. In a concrete interpretation, the duration of a variation lies between eight and twelve seconds. For the notation of the musicians' interpretations in real-time, Bel developed an application which enables one to type stenographic notes of bols on the keyboard of an Apple II. An additional extension of this word processor in terms of search and substitution strategies finally leads to the implementation of an inference module (named QAVAID) for the processing of generative grammars. In order to generalize this model from the musicians' improvisations, finite automata are produced whose terminals comprise of either single bols or a number of them. Figure 4.13 illustrates an example of such a finite automaton; the symbols marked with X(x) indicate the nodes of the graph that may branch into different formal segments. A hyphen indicates a pause in the length of a bol; in this specific case, however, the duration of the precedent bol is doubled. Figure 4.14 shows an alternative notation within a two-layer model in which the respective bols and their combinations on the edges are marked as terminal symbols T(x); the number after the corresponding indication of the edges refers to the amount of rhythmic units.

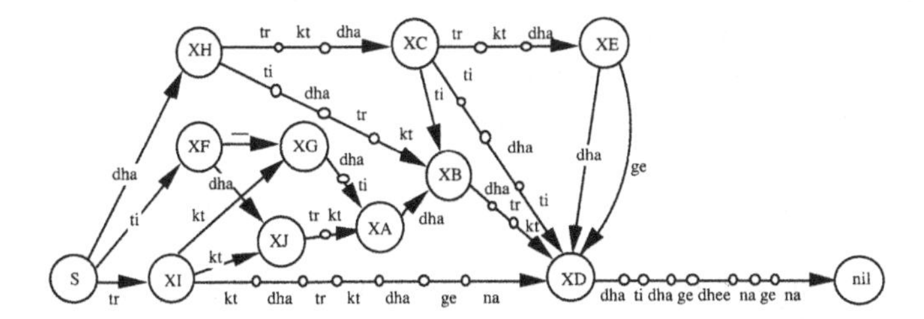

Fig. 4.13 Finite automaton for the representation of bol sequences. With kind permission of Bernard Bel.

The major difficulty with this system lies in the generation of a musical terminal alphabet, i.e. the coherent segmentation of the bol sequences into "words" in order to produce an implied rhythmic "vocabulary" of the tabla improvisations. Although the finite automaton shown above is indeed able to generate some correct examples,

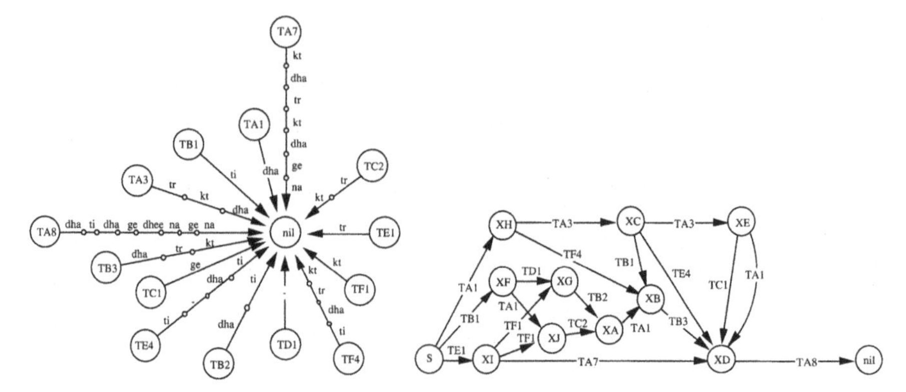

Fig. 4.14 Two-layer model of the finite automaton from figure 4.13. With kind permission of Bernard Bel.

the segmentation of bol sequences is, however, unsatisfactory in some cases: "[. . .] chunks like *trkt*, *dhatrkt*, and *dhatidhagedheenagena* may be called 'words' or 'sequences of words' in the sense that they represent blocks that can be substituted or permuted. However, a unit like *kt* is never used as a separate block but is always preceded by *tr* [. . .]." [5, p. 10]. In order to resolve this problem and consequently to improve the ability of the software to generalize, QAVAID is extended to an adaptive system whose output is evaluated by a user and improved using backtracking mechanisms in case the solutions are insufficient. Tables 4.2 and 4.3 show a type-3 grammar produced by QAVAID which, due to the preceding learning process, also exhibits a considerably better segmentation of the bol sequences (the numbers after the variables S(x) and the terminals T(x) denote the amount of the contained bols).

Regardless of its analytical potential, the Bol Processor represents at present a powerful tool for working with generative grammars in the field of algorithmic composition. The software enables the processing of both note values and sound objects as terminal symbols. In the BP, a number of functions considerably extend the range of possible applications within a traditional generative grammar. In this context, for instance, rewriting rules may process *wild cards*[13] or also involve a broader context, meaning strings of symbols on any position before or after the actual substitution. Additionally, in the selection of rewriting rules, different probabilities may be determined for their application and preferences for the order of their processing may be defined.

A particular aspect of the Bol Processor can be found in the representation of musical time structure. In the representation of time, fragments of musical structure are related to one another. Two principle operators determine whether the musical fragments are played simultaneously or sequentially. As an example, three musical structures, A, B and C are established that consist, respectively, of one, two or three notes of equal duration initially. When an operator for equal relation of duration is set between the three units, all three structures receive, in relation to one another,

[13] Here: Placeholder for arbitrary terminal or non-terminal symbols.

GRAM#1

S	→	TA3	SA13
SA13	→	TA3	SA10
SA10	→	TC2	SA8
SA8	→	TD2	SA6
SA6	→	TA6	
SA10	→	TA2	SA8
SA13	→	TC5	SA8
SA13	→	TD2	SB11
SB11	→	TA3	SA8
S	→	TB2	SA14
SA14	→	TD2	SA12
SA12	→	TA2	SB10
SB10	→	TB2	SA8
S	→	TE2	SA14
S	→	TF2	SB14
SB14	→	TB2	SA12
S	→	TB2	SB14
...			
S	→	TB2	SC14
SC14	→	TA3	SA11
SA11	→	TB3	SA8
S	→	TD2	SD14
SD14	→	TA3	SB11

Table 4.2 Type-3 grammar generated by QAVAID, part 1.

GRAM#2

TB3	→	dhagena
TF2	→	tidha
TE2	→	ti-
TC5	→	dhati-dhati
TA6	→	dhagedheenagena
TD2	→	dhati
TC2	→	dhage
TA3	→	dhatrkt
TB2	→	trkt
TA2	→	dhadha

Table 4.3 Type-3 grammar generated by QAVAID, part 2.

the same duration like, for example, a quarter (A), two eighth notes (B) and eighth triplets (C). Now, if, for example, an operator indicating simultaneity is set after A, one voice plays A, the second voice plays B and C successively, where B and C together have the same duration as A, e.g. a half note (A), two eighth notes (B) and eighth triplets (C). This syntax enables an approach to metric concepts of non-

Western musical traditions that are hardly realizable through traditional notation systems.[14]

Furthermore, the Bol Processor is able to integrate "performance rules" that enable a differentiated time structure for the playback of the composition: "When dealing with notes of discrete sound-objects in computer music, the traditional approach was to produce a representation of the musical work whose timings could be mapped to Western staff notation. To play the work, a human would follow instructions written on top of the score: *rallentando*, *rubato*, etc., and a machine would need interpretation rules generally stochastic and restricted to local variations of the tempo. In BP the 'score' is not the final musical work. It does not contain the timings of all phrases, but rather a hierarchy of rhythmic patterns that may be as complex as required by the composition. This information is further processed to produce the final timings. The difference is striking as the performance may sound very smooth even though no stochastic rules have been used. The advantage of this approach is that the mechanical performance retains a consistency that only skilled human interpreters would be able to render."[15] One of the advantages of using systems of algorithmic composition that allow for the application of generative grammars can be found, furthermore, in the possibility of producing complex structures before beginning the information processing: "I think that the development of more and more visual stuff curtails the possibility of 'thinking in your chair.' Sometimes I develop grammars, not at the computer, but sitting with pencil and paper. With programs [other than BP2] this is not possible: you must sit in front of the computer. The difference lies in the type of attention that each software environment demands on the part of the composer, and indeed reflects on the way s/he thinks about music."[16]

4.2.5 *Jazz*

Mark Steedman [41] described a grammar for the generation of jazz chord sequences. The output of the system consists of twelve-bar blues progressions that are generated by rewriting rules from a simple initial phrase (rule 0):

The expressions in brackets on the left side of the production rule are alternative forms of the chords that must be substituted in the same way on the right side of the rule (figure 4.15). A subscript non-terminal (x) denotes the chord on which the actual harmonic function is applied. A "w" stands for the chord that is subject to substitution that for example in rule 3b becomes a dominant seventh chord of the successor and the round brackets in rule 6 indicate alternative chords. Figure 4.16

[14] For pulse and quantization in "non-Western" music, see [10].

[15] Comment by Bel in an E-mail to the author.

[16] Composer Harm Visser on his motivation to use the BP, cf. [11, p. 10].

0:			S12(m) →	I(m)	I7	IV(m)	I(m)	V7 I(m)	
1:			x(m)(7) →	x(m)		x(m)(7)			
2:			x(m)(7) →	x(m)(7)		Sd_x			
3a:		w	x7 →	D_x(m)7		x7			
3b:		w	xm7 →	D_x7		xm7			
4:		D_x7	x(m)(7) →	♭St_x(m)(7)		x(m)(7)			
5:	x	x	x →	x		St_xm	M_xm		
6:	x(m)	x(m)	{ D_x / St_xm7 / L_xm7 } →	x(m)		♯x°7	{ D_x / St_xm / L_xm7 }		

Fig. 4.15 Rewriting rules for the generation of twelve-bar phrases [41, p. 68]. With kind permission of Mark Steedman.

shows a successive application of the rules 1 to 6 resulting in a particular chord progression.[17]

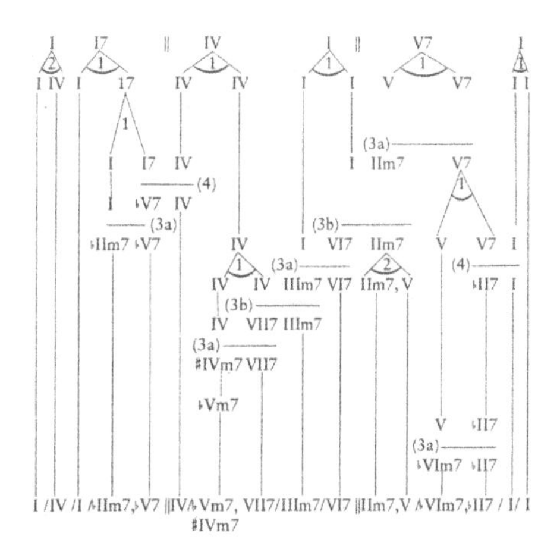

Fig. 4.16 Chord sequences by rewriting rules [41, p. 71]. With kind permission of Mark Steedman.

Philip N. Johnson-Laird [24] dealt with structural aspects of jazz improvisations, also by applying generative grammars, which establish simple constructive principles for rhythmic structures, chord progressions and bass lines. In regard to rhythm, a regular grammar produces two-bar phrases, as can be seen in figure 4.17.

Based on the investigation of blues progressions, a number of context-free rules that are divided into hierarchical classes [24, p. 311–312] generate eight-bar tonal chord sequences (figure 4.18).

[17] In his work, Mark Chemellier describes some interesting extensions to Steedman's model which amongst other things allow for a variable setting of the depth of the substitution process as well as influencing the choice of the various chord substitutions, cf. [18].

Bar 1:			Bar 2:		
Beat-1	→ ♩	Beat-2	Beat-1	→ ♩	Beat-2
	♩.	Beat-2.5		→ ♩.	Beat-2.5
Beat-2	→ ♪	Beat-2.5		→ 𝅗𝅥	Beat-3′
	♩	Beat-3	Beat-2	→ ♩	Beat-3
Beat-2.5	→ ♪	Beat-3′		→ ♩	Beat-3′
Beat-3	→ ♪	Beat-3.5	Beat-2.5	→ ♪	Beat-3′
	♩	Beat-4	Beat-3	→ ♩	Beat-4
	♩.	Beat-4.5	Beat-3′	→ 𝅗𝅥	Bar 3...
Beat-3′	→ ♪	Beat-3.5	Beat-4	→ ♩	Bar 3...
	♩	Beat-4			
Beat-3.5	→ ♪	Beat-4			
Beat-4	→ ♩	Bar-2 Beat-1			
Beat-4.5	→ ♪	Bar-2 Beat-1			

Fig. 4.17 Regular grammar for two-bar rhythmic models [24, p. 303]. With kind permission of Philip Johnson-Laird.

Eight-bars	→	First-four Second-four		
First-four	→	Opening-cadence		Opening-cadence
	→	Opening-cadence′		Opening-cadence
Second-four	→	Middle-cadence		Opening-cadence
Opening-cadence	→	| I	| I	|
	→	| I	| V	|
Opening-cadence′	→	| I	| III	|
	→	| I	| IV	|
Middle-cadence	→	| I	| IV	|
	→	| I	| V	|
	→	| IV	| I	|

I	IV	I	V	IV	I	I	V
I	III	I	V	I	IV	I	I
I	IV	I	V	I	IV	I	I
I	I	I	I	IV	I	I	I
I	V	I	V	I	V	I	I

Fig. 4.18 Context-free grammar for eight-bar chord sequences and example generation [24, p. 310]. With kind permission of Philip Johnson-Laird.

The bass lines are generated by a twofold process, where first a melodic contour is produced by a regular grammar to which the concrete pitches from the previously generated chord progressions are assigned. Here, 1st, 3rd and 7th chord notes are preferred and passing notes are avoided for the beginning of a new chord and on the first beat of the bar. Figure 4.19 shows a graph of the grammar for the melodic contour and one resulting bass line ("f" here means the starting note of each bar, "d" indicates a repetition of the preceding note, "s" is a second progression, and "i" is every interval higher than a second).

Further parameters such as timbre, volume and articulation are not taken into consideration for the grammatical modeling, which is intended to provide a skeletal structure out of which, similar to a Lead sheet, the actual musical structure has to be developed.

François Pachet [32] used different approaches for the description of musical structure, an essential aspect of his work being the examination of musical surprise: "Most of the works in music cognition relate surprise to the phenomenon of musical

Fig. 4.19 Melodic contour and possible bass line [24, p. 316–317]. With kind permission of Philip Johnson-Laird.

expectation. [...] In this paper, we emphasize the importance of the rich algebraic structure underlying Jazz chord sequences, and suggest that harmonic surprise may not only be related to unexpected structures, but also to 'calculus', i.e. to an ability to deduce a sequence from a set of combinatorial rules." [32, p. 1]. By means of a data compression method, chord sequences are analyzed and represented. This leads to their classification into expected and surprising harmonic progressions. Furthermore, stochastic procedures are applied to deduce rewriting rules from the given corpus. Chord progressions, as shown in table 4.4, are used as a basis for further examinations. By means of the analysis of frequently recurring harmonic movements,

Blues For Alice
F | E halfDim7 / A7 | D min / G7 | Cmin / F7 | Bb7 |
Bbmin / Eb7 | Amin | Ab min / Db7 | Gmin7 | C7 | F7 | Gmin / C7 |

Marmaduke
Gmin | Gmin | Gmin | Gmin / C7 | F | Gmin / C7 |
F | Amin / D7 | Gmin | Gmin | Gmin | Gmin / C7 | F |
Gmin/ C7 | F | F |Cmin | F7 | Bb | Bb | G7 | G7 |Gmin |
C7 | Gmin | Gmin | Gmin | Gim / C7 | F |
Gmin / C7 | F | Amin / D7 |

Nows The Time
F7 | F7 | F7 | F7 | Bb7 | Bb7 | F7 | D7 |
Gmin | C7 | F7 | C7 |

Ornithology
G | G | Gmin | C7 | F | F | Fmin | Bb7 | Eb7 |
A halfDim7 / D7 | Gmin | D7aug9 | Bmin | E7 |
Amin | D7 | G | G | Gmin | C7 | F | F | Fmin |
Bb7 | Eb7 | AhalfDim7 / D7 | G | G | Bmin / E7 |
Amin / D7 | G / E7 | Amin / D7 |

Table 4.4 Chord sequences used by Pachet.

such as II-V-I or I-VI-II-V cadences, as well as further rules of jazz harmony, nine classes of rewriting rules are formulated:

1. Repetition: C → C / C (C and C → C take the same amount of time; "/" separates second and third counting time)
2. Enrichment of chords: C7 → C7^9 (for all additionally possible chord notes)
3. Relative minor: C → A min
4. Tritone Substitution: C7 → F#7
5. Preparation: C → G7/C
6. Preparation by Minor Seventh: C7 → Gmin7/C7
7. Transition to Fourth: C7 →C7/F
8. Back propagation of Seventh: XXC7Y → XC7YY (At the same time substitution of the free position by the subsequent chord)
9. Left deletion: XC7 → XX (The deletion of chords forms a class of rewriting rules whose concrete application must be decided according to the respective harmonic context)

Table 4.5 shows an example of a successive application of the rewriting rules on the C7 chord in the fourth bar: Preparation by Minor Seventh (6), Tritone Substitution (4), Preparation (5), Back Propagation of the former chord (8) and deletion (9). Pachet is not only interested in "well-formed," rule-compliant chord sequences, but

C | F | C | C7 | F ...
C | F | C | Gmin7 / C7 | F ...
C | F | C | Gmin7 / F#7 | F ...
C | F | C | D7 / Gmin7 / F#7 | F
C | F | C / D7 | Gmin7 / F#7 | F
C | F | D7 | Gmin7 / F#7 | F ...

Table 4.5 Stepwise chord substitution in a simple cadence.

also aims at establishing a criterion for "surprise" and "expectation." For this purpose, he represents the chord in the context of a data compression procedure [46]. The representation of the data referred to as the *Lempel–Ziv tree* (*LZ-tree*) always only encodes the shortest symbol strings that have not yet been identified. Representation is made within a directed graph. As an example (figure 4.20), a symbol string abcabacabbbbbba is encoded in segments indicating the sequences of symbols as paths from the root. This procedure deconstructs the symbol string in the segments a b c ab ac abb bb bba and builds up an appropriate tree structure.

Before the Lempel–Ziv encoding, the chord sequences are represented by the chord structure and the interval relation between the roots. For the Lempel–Ziv encoding, the chord progressions are simplified by indicating every chord transition from C, as shown in table 4.6 by means of a short cadence.

The representation within the Lempel–Ziv tree shows the frequency of particular chord progressions. Those movements that occur very rarely are considered "surprising." A progression, represented by a particular node, occurs in the encoded

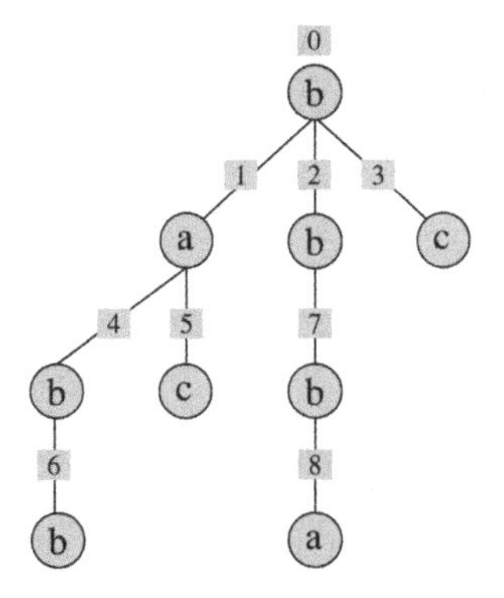

Fig. 4.20 Lempel–Ziv tree representing a symbol string.

E	Amaj7	F#min7	B7
	(C : Fmaj7)	(Cmaj7 : Amin7)	(Cmin7 : F7)

Table 4.6 Transition of a chord sequence.

corpus more rarely the fewer children it has. As an example, figure 4.21 shows a segment of the LZ–tree built from the corpus. If the preceding sequence Cmin - F7 - B^b is encoded as (Cmin) - (F7=C7) - (F=C), the most surprising succeeding chord is (Emin=Cmin), presented as a framed node.

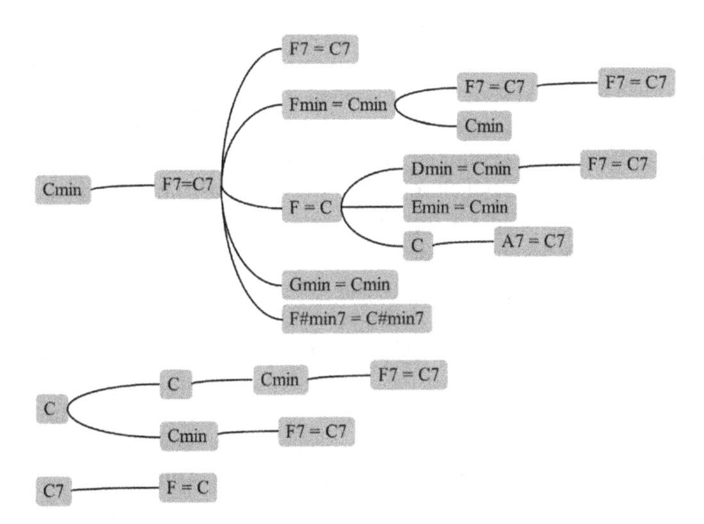

Fig. 4.21 Segment of the Lempel-Ziv tree of the chord sequences used by Pachet.

By means of this procedure, Pachet demonstrates a method of encoding musical tasks with generative models. The two examined aspects in this case, "expectation" and "surprise," are examples of an interesting "measuring" of stylistic features by the extension of traditional methods of generative grammars. Another approach in Pachet's work treats the automatic derivation from a given corpus by statistical methods, as described in the following chapter.

4.2.6 Grammatical Inference

Pachet's method allows for the automatic recognition of style-compliant chord substitutions due to a given comprehensive corpus. The underlying principle first generates the set of all possible rewriting rules and consequently comprehends both common as well as unusual chord substitutions. In a further step, each of these rules is examined for style compliance by applying them to the corpus and comparing the distribution of neighbors before and after the substitution. To do this, sequences of each three consecutive chords are compared, the second chord of the sequence representing the beginning of the next group and so forth. For a sequence T S T S D7 etc., the following groups of three would result: T S T, S T S, T S D7, etc. This principle allows each chord to be comprehended with its context in both corpora. A good chord sequence is characterized by the fact that "chord new" – before applying the substitution – stands in the same contexts as "chord old" that in a next step is substituted by "chord new." If, for example, a dominant seventh chord (D7) is substituted by a seventh and ninth chord (D7/9), in a corpus of sufficient size, the D7/9 can already be found before the substitution in contexts, where the D7 also appears. So, the similarity of the distribution of neighbors in the corpus before and after a particular chord substitution turns out to be an interesting indicator for its style compliance.

Craig Nevill-Manning and Ian Witten [31] described the automated generation of rewriting rules based on an input of symbol strings that were produced by a Lindenmayer system; furthermore, the authors broaden their approach in regard to linguistic constructs of a larger extent and chorale melodies composed by J.S. Bach. An entered symbol string is examined for repeated combinations of symbols, for which then a non-terminal symbol is set. In order to guarantee an efficient procedure, two conditions are introduced: Each pair of adjacent symbols appears only once in a production rule ("digram uniqueness") and in the generation of the symbol string, every production rule is used more than once. This ensures that the rule is useful as a structure-determining procedure in contrast to a literal depiction of every similar symbol string which occurs multiple times ("rule utility"). In order to illustrate this principle, figure 4.22 shows the generation of production rules by Nevill-Manning and Witten's software SEQUITUR, on the basis of a given symbol string.

In the sequential input of new symbols, the system permanently runs through a multi-stage process. New terminals are added to the right side of the production rules until repetition occurs. In this case, the respective symbol string is replaced by a non-terminal symbol. When the software is applied for the analysis of large text passages, good results can be achieved regarding the segmentation of units belonging syntactically together. In the analysis of soprano voices in Bach chorales,

symbol number	the string so far	resulting grammar	remarks
1	a	S → a	
2	ab	S → ab	
3	abc	S → abc	
4	abcd	S → abcd	
5	abcdb	S → abcdb	
6	abcdbc	S → abcdbc	bc appears twice
		S → aAdA A → bc	enforce digram uniqueness
7	abcdbca	S → aAdAa A → bc	
8	abcdbcab	S → aAdAab A → bc	
9	abcdbcabc	S → aAdAabc A → bc	bc appears twice
		S → aAdAaA A → bc	enforce digram uniqueness. aA appears twice
		S → BdAB A → bc B → aA	enforce digram uniqueness
10	abcdbcabcd	S → BdABd A → bc B → aA	Bd appears twice
		S → CAC A → bc B → aA C → Bd	enforce digram uniqueness. B is only used once
		S → CAC A → bc C → aAd	enforce rule utility

Fig. 4.22 Grammatical inference in SEQUITUR [31, p. 70]. With kind permission of Craig Nevill-Manning.

SEQUITUR correctly identifies repeated melodic phrases as well as final cadences [31, p. 72–73].

Teuvo Kohonen [25] presented an interesting algorithm of grammatical inference. As a Self-Learning Musical Grammar, [26] his algorithm enables the recognition of a context of variable length within a musical corpus and generates a context-free grammar for the production of new melodic material.

The production rules of a generative grammar should be able to clearly describe and regenerate a corpus. In order to achieve this, each successor must be defined uniquely by its preceding context; the context depth necessary for uniqueness may be different for different symbols. If, for example [26], on the basis of a symbol string ABCDEFG ... IKFH ... LEFJ ... a concrete successor of F should be deduced, different possibilities of applying the production rules result depending on the respective context level. In context level 1, the successors are G, H and J; therefore, it is not possible to determine an exact successor solely on the basis of F. If the context level is now extended to two symbols (EF, KF, EF), then KF is uniquely followed by H and EF has the alternative successors G and J. Only at context level 3 (DEF, LEF) may the definite successors G (context: DEF) and J (context: LEF) be detected. This example shows that for the production of different successor symbols in a corpus, different context levels are required – this is a condition that, for example, cannot be modeled by a Markov model due to the previously set order, i.e. context level. Kohonen's approach represents a sort of Markov model of variable order which formulates, for each symbol of the input, the required context levels – the number of preceding symbols in the form of production rules that are able to describe exhaustively the order of symbols in the corpus. The algorithm reads the symbols of the input successively and produces rules of the type predecessor → successor. If a preceding symbol is again available as an input and determines another successor, the rule generated before is provided with a so-called "conflict bit" and the context is extended until no conflict occurs any longer in the application of the rules, e.g. KF → H, DEF → G, and LEF → J. The generated rules may also be represented by a tree structure, as shown in figure 4.23.

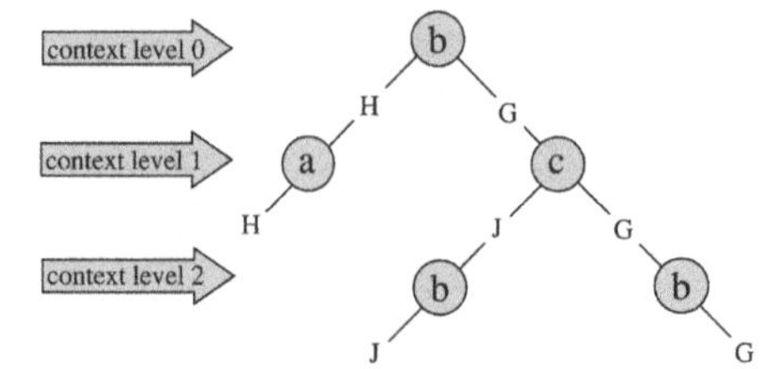

Fig. 4.23 Tree with variable context levels.

For the generation of new material, a "depth parameter" can be set which determines how many nodes are subtracted from the "optimal" context of each symbol. So, rules are applied that previously had been provided with a conflict bit – in the above case, for example, given a depth parameter of 1, symbol sequences of the form DEF → J or LEF → G may also be generated that are not contained in the sequence of symbols of the input. As an expansion of this coherent model, Kohonen described different encoding strategies for harmonic information as well as possible extensions of the system [26, p. 4].

As another possibility of processing a variable context, I have designed a simple algorithm which is very easy to implement and provides another possibility to build a structure based on the different context depths within a corpus. In contrast to Kohonen's Self-Learning Musical Grammar, no rewriting rules are produced here, but the algorithm passes through regions of examples of a given corpus, where the trace is forming the final output at the same time. This algorithm is initialized with a symbol or a sequence of symbols as a predecessor and detects the possible successor symbols in the corpus. In the next step, one of these successors is randomly selected and added to the predecessor to form a new predecessor, and the process repeats until only a single possible successor is left. Through this, the context depth is extended as long as there is only one possible successor left. By means of backtracking, the context depth is now decreased until again alternative possibilities for continuing the process become available. The scheme below illustrates how the algorithm works. In this example, the symbols denote musical terminals, which can encode a single or a number of parameters. The musical examples of the corpus may be copied one after the other in a list; herein, it is recommended to mark the terminals that are positioned at the beginning and the end of each example. The indication of the end of each example enables the algorithm to stop the generation and to output the result. While marking the beginning symbols is not strictly necessary, it allows for starting with a predecessor which is actually used as the beginning symbol of an example of the corpus.

1. Set a predecessor (P) at which the search should start; P becomes the first element of the list "Result" (R).
2. Produce an empty list "Alternatives" (A).
3. Determine all successors (S) of P from the corpus.
   Every $P+S$ becomes a new P.

Every P becomes an element of A.
If number of elements in $A = 1$, continue with point 4.
If number of elements in $A > 1$, continue with point 3.

4. Choose a P from A randomly and pass the last symbol of this P to R. If last symbol of P = symbol for the end of a segment, continue with point 5. Otherwise: Continue with point 2.
5. If last symbol of P = symbol for the end of a segment, pass this symbol to R, continue with point 5. Otherwise: Delete first symbol of P and continue with point 2.
6. Output R / END.

Since this algorithm behaves similarly to a snake which devours more and more material, but insists – in proper gourmet style – on a selection of at least two alternative bits, this procedure may well be called a "Context Snake."[18] As an extension of this model, a "backtracking parameter" is applied which represents a kind of lower bound for a particular context depth. The parameter determines the minimal necessary number of symbols of the predecessor. In case the number is below that, alternative branches are selected recurrently until – except, of course, at the beginning – each transition of the new generation shows the minimal necessary context depth. This "backtracking parameter" allows for precisely adjusting to which extent the new generations reflect the contextual characteristics of the corpus.

Experiments with this algorithm yielded good results in regard to the generation of chorale melodies. For the corpus, Dorian melodies were used that conform to rules according to Knud Jeppesen. In his study book of classic vocal polyphony [23], this Danish composer and musicologist described methods for the "composition" of contrapuntal movements with regard to the style of Pierluigi Palestrina (around 1525–1594). Jeppesen divides contrapuntal material into rhythmic variants as well as movements of different numbers of voices and determines precise conditions and rules for a style-compliant generation of each of these variants. Here, monophonic melodies of the first "species" describe note sequences of equal rhythmic duration that may form the basis, known as *cantus firmus*, of a polyphonic movement. Because Jeppesen's rules are well suited to examine the melodic material for its style conformity, reasonably objective evaluations of generations of this non-knowledge-based system are possible. Figure 4.24 shows an example for the passing of the "Context Snake" through a corpus of Dorian melodies: The light grey areas indicate the processed context depths and the darker grey areas indicate the trace resulting in the output, as shown in the last two staves. A value of 3 was set for the "backtracking parameter" which results in a generation comparable to a "variable" markov model of third and higher order.

Constraints prove to be reasonable extensions for the examination of the generated material in terms of length, forbidden frame intervals and inappropriate repetitions. In a small corpus, it is also possible to re-generate original examples, because

[18] The "Context Snake" has been realized as a pattern for the language SuperCollider by Alberto de Campo. It is availabe as an extension package at http: supercollider.sf.net.

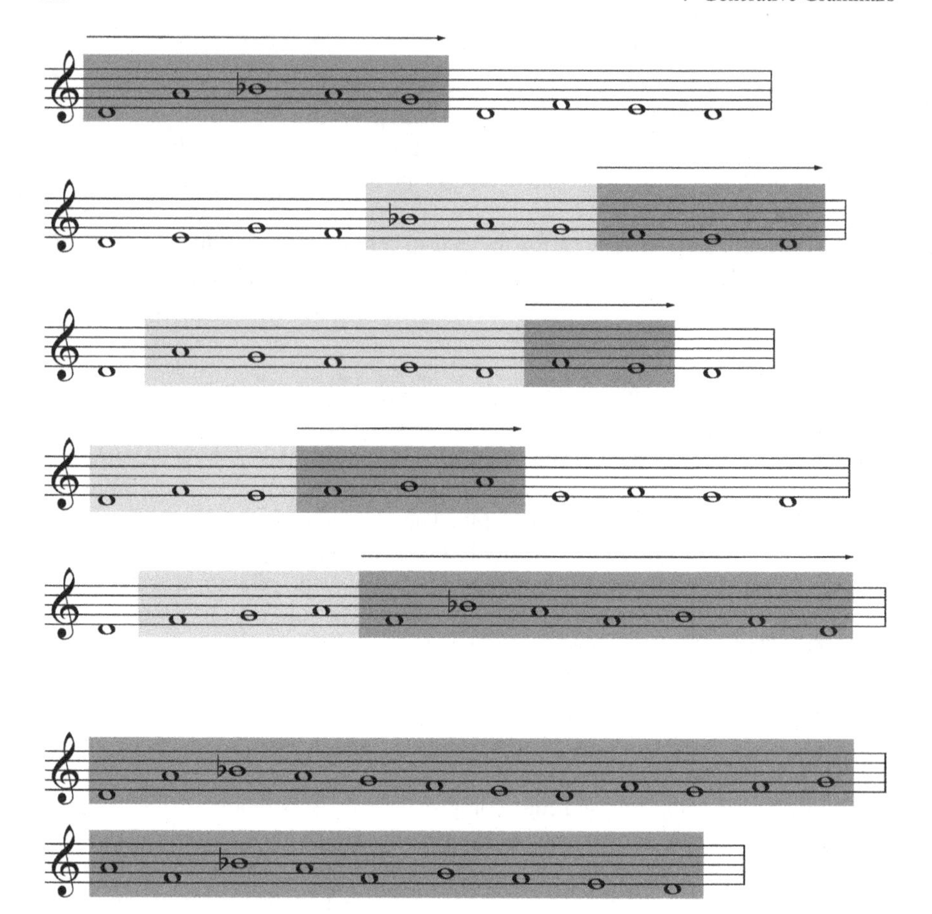

Fig. 4.24 Context Snake passing through a corpus of Dorian melodies.

here the algorithm repeatedly selects branches by chance that may finally produce an example of the corpus.

Despite these restrictions, interesting musical results are also generated within musical styles other than chorale melodies. Furthermore, it proves to be advantageous that the output reflects the distribution of the transition probabilities in the corpus, since predecessors occurring frequently are selected with a higher probability. However, as for each new generation in most cases an exhaustive backtracking is applied, this algorithm requires longer calculation periods.

An exhaustive analysis model that, for the generation of a grammar, also determines the semantic meaning of musical terminals on different hierarchical levels is described by David Cope, who produces style imitations of musical genres with his system EMI (see chapter 5).

4.3 Synopsis

Generative grammars are an essential class of formalisms of musical analysis in algorithmic composition which allow for a context-related and hierarchical organization of musical material. However, a significant restriction which this formalism shares with other methods coming from linguistics is their basic orientation on a sequential model that must neglect simultaneously occurring structure – a feature which is naturally of the utmost importance in polyphonic music.

As in Markov models, transition networks or Lindenmayer systems, extended strategies for the structural treatment of coexisting musical layers must be developed. Due to this difficulty, generative grammars are often used for the modeling of musical styles that lend themselves to a sequential treatment of the material, such as for monodic melodies or chord progressions of different stylistic provenience.

The possibility of the grammatical transformation of linguistic expressions, as it is used by Chomsky for the representation of different word orders of the same semantic content, referred to as "deep structure," is hardly applicable to a musical context. Where different positions of constituents in a linguistic expression may, however, show the same semantic meaning, the reorganization of musical segments generally leads to new musical information. In addition, the term "semantics" can hardly be applied to music in general, because in this case the "meaning" cannot be distinguished from the correct appearance of the musical structure. However, within Schenker's approach and similar methods, the term "deep structure" is often used to refer to a skeleton of musical structure that rather represents a construction principle and is incomplete regarding its musical appearance.

A decisive aspect of generative grammar lies in its high generative capacity and the possibility of generating a complex musical structure by using only a comparatively small number of rewriting rules. In the area of style imitation, some approaches of grammatical inference offer a great advantage through the possibility of processing a variable context depth, but nevertheless it should not be overlooked that the "well-formedness" of a generated expression does not present a guarantee of a satisfying musical structure. Moreover, a number of musically determining factors are often ignored in the generation of grammatical models and also simplifications are often made, leading in the worst case to the arbitrariness of the generated structure.

In contrast to, for example, genetic algorithms or cellular automata where a permanent flow of musical material is produced, musical structure in a generative grammar does not result until the end of parsing is reached, where the preceding substitution processes cannot be used for an actual musical output due to the residual non-terminal symbols. This aspect of generative grammar, however, turns out to be a disadvantage only for concepts that deliberately intend to work with a process-like character. On the whole, due to the wide spectrum of possible applications, generative grammars represent a very useful and versatile class of algorithms for musical analysis, style imitation and genuine composition.

References

1. Ahlbäck S (2004) Melody beyond notes: a study of melody cognition. Skrifter från Institutionen för Musikvetenskap, 77. Göteborgs Universitet, Göteborg. ISBN 91 8597473-0
2. Baroni M, Brunetti R, Callegari L, Jacoboni C (1982) A grammar for melody. Relationships between melody and harmony. In: Baroni M, Callegari L (eds) Musical grammars and computer analysis. Casa Editrice Leo S. Olschki, Florence. ISBN 2147483647
3. Baily J (1989) Principles of rhythmic improvisation for the Afghan rubâb. Intl. Council for Traditional Music UK Chapter Bulletin, 1989
4. Becker A, Becker JO (1979) A Grammar of the musical genre Srepegan. Journal of Music Theory, 23, 1979
5. Bel B, Kippen Jim (1989) The identification and modelling of a percussion "language," and the emergence of musical concepts in a machine-learning experimental setup. Computers and the Humanities, 23/3, 1989
6. Bel B, Kippen J (1992) Bol Processor Grammars. In: Balaban M, Ebcioglu K, Laske O (eds) Understanding music with AI. AAAI Press/MIT, Cambridge, Mass. ISBN 0262-52170-9
7. Bel B, Kippen J (1992) Modeling music with grammars. In: Marsden A, Pople A (eds) Computer representations and models in music. Academic Press, London. ISBN 0-12-473545-2
8. Bel B (1992) Symbolic and sonic representations of sound-object structures. In: Balaban M, Ebcioglu K, Laske O (eds) Understanding music with AI. AAAI Press/MIT, Cambridge, Mass. ISBN 0262-52170-9
9. Bel B (1996) A flexible environment for music composition in non-European contexts. Journées d'Informatique Musicale 1996, Caen (France)
10. Bel B (1996) A symbolic-numeric approach to quantization in music. 3rd Symposium on Computer Music, Recife (Brazil), 1996
11. Bel B (1998) Migrating musical concepts – an overwiew of the Bol Processor. Computer Music Journal, 22/2, 1998
12. Blacking J (1970) Tonal organisation in the music of two Venda initiation schools. Ethnomusicology, 14, 1970
13. Blacking J (1982) What languages do musical grammars describe? In: Baroni M, Callegari L (eds) Musical grammars and computer analysis. Casa Editrice Leo S. Olschki, Florence. ISBN 2147483647
14. Bloomfield L (1933) Language. Rinehart and Winston, New York. Reprinted by University of Chicago Press, 1984. ISBN 0226060675
15. Camillieri L (1982) A grammar of the melodies of Schuberts Lieder. In: Baroni M, Callegari L (eds) Musical grammars and computer analysis. Casa Editrice Leo S. Olschki, Florence. ISBN 2147483647
16. Chomsky N (1957) Syntactic structures. Mouton, Den Haag. Reprinted by Walter de Gruyter, Berlin, New York, 1989. ISBN 9027933855
17. Chomsky N (1965) Aspects of the theory of syntax. MIT Press, Cambridge, Mass. ISBN 0262530074
18. Chemellier M (2001) Improvising jazz chord sequences by means of formal grammars. http://recherche.ircam.fr/equipes/repmus/marc/publi/jim2001/icmc2001.pdf Cited 1 Feb 2005
19. Fanselow G, Felix SW (1993) Sprachtheorie. Eine Einführung in die generative Grammatik, 2: Die Rektions- und Bindungstheorie, 3rd edn. Francke, Tübingen. ISBN 3-7720-1732-0
20. Deliège C (1982) Some unsolved problems in Schenkerian theory. In: Baroni M, Callegari L (eds) Musical grammars and computer analysis. Casa Editrice Leo S. Olschki, Florence. ISBN 2147483647
21. Hausser R (2000) Grundlagen der Computerlinguistik. Springer, Berlin, Heidelberg. ISBN 3-540-67187-0
22. Hughes DW (1991) Grammars of non-Western music. In: Howell P, West R, Cross I (eds) Representing musical structure. Academic Press, London. ISBN 0-12-337171-5

23. Jeppesen K (1992) Counterpoint: the polyphonic vocal style of the sixteenth century. Dover Publications, New York. ISBN 04862736X
24. Johnson-Laird PN (1991) Jazz Improvisation: a theory at the computational level. In: Howell P, West R, Cross I (eds) Representing musical structure. Academic Press, London. ISBN 0-12-337171-5
25. Kohonen T (1987) Self-learning inference rules by dynamically expanding context. In: Proceedings of the IEEE First Annual International Conference on Neural Networks, San Diego, 1987
26. Kohonen T (1989) A self-learning musical grammar, or "Associative memory of the second kind." In: Proceedings of the International Joint Conference on Neural Networks, New York, 1989
27. Lerdahl F, Jackendoff R (1983) A generative theory of tonal music. MIT Press, Cambridge, Mass. ISBN 0-262-62107-X
28. Lerdahl F, Jackendoff R (1983) An overview of hierarchical structure in music. In: Schwanauer SM, Levitt DA (eds) Machine models of music. MIT Press, Cambridge, Mass. ISBN 0-262-19319-1
29. Lerdahl F (1991) Underlying musical schemata. In: Howell P, West R, Cross I (eds) Representing musical structure. Academic Press, London, pp 273–290. ISBN 0-12-337171-5
30. Meehan JR (1979) An artificial intelligence approach to tonal music theory. In: Martin AL, Elshoff JL (eds) Proceedings of the 1979 annual conference. Association for Computing Machinery, New York, pp 116–120
31. Nevill-Manning CG, Witten IH (1997) Identifying hierarchical structure in sequences: A linear-time algorithm. Journal of Artificial Intelligence Research, 7, 1997. pp 67–82
32. Pachet F (1999) Surprising harmonies. International Journal of Computing Anticipatory Systems, 1999
33. Pardo B, Birmingham WP (2000) Automated partitioning of tonal music. In: Etheredge JN, Manaris (eds) Proceedings of the Thirteenth International Florida Artificial Intelligence Research Society Conference. AAAI Press, Menlo Park, Calif, pp 23–27
34. Pelinski R (1982) A generative grammar of personal Eskimo songs. In: Baroni M, Callegari L (eds) Musical grammars and computer analysis. Casa Editrice Leo S. Olschki, Florence. ISBN 2147483647
35. Rader GM (1974) A method for composing simple traditional music by computer. In: Schwanauer SM, Levitt DA (eds) Machine models of music. MIT Press, Cambridge, Mass. ISBN 0-262-19319-1
36. Roads C (1985) Grammars as representations for music. In: Roads C, Strawn J (eds) Foundations of computer music. MIT Press, Cambridge, Mass. ISBN 0-262-18114-2
37. Roads C (1982) An overview of music representations. In: Baroni M, Callegari L (eds) Musical grammars and computer analysis. Casa Editrice Leo S. Olschki, Florence. ISBN 2147483647
38. Schenker H (1935) Der freie Satz, 2nd edn. Neue musikalische Theorien und Phantasien, 3. Universal-Edition, Vienna. ISBN B0000BN9B5
39. Schöning U (2003) Theoretische Informatik – kurzgefasst, 4th edn. Spektrum Akademischer Verlag, Heidelberg, Berlin. ISBN 3-8274-1099-1
40. Smoliar SW (1979) A computer aid for Schenkerian analysis. In: Martin AL, Elshoff JL (eds) Proceedings of the 1979 annual conference. Association for Computing Machinery, New York, pp 110–115
41. Steedman M (1984) A generative grammar for jazz chord sequences. Music Perception 2, 1984
42. Steedman M (2003) Formal grammars for computational musical analysis. INFORMS Atlanta October 2003
43. Sundberg J, Lindblom B (1976) Generative theories in language and music descriptions. In: Schwanauer SM, Levitt DA (eds) Machine models of music. MIT Press, Cambridge, Mass. ISBN 0-262-19319-1

44. Sundberg J, Lindblom B (1991) Generative theories for describing musical structure. In: Howell P, West R, Cross I (eds) Representing musical structure. Academic Press, London. ISBN 0-12-337171-5
45. Temperley D, Sleator D (1999) Modeling meter and harmony: a preference-rule approach. Computer Music Journal 23/1, 1999
46. Ziv J, Lempel A (1978) Compression of individual sequences via variable-rate coding. IEEE Transactions on Information Theory, 24/5, September, 1978

Chapter 5
Transition Networks

Transition networks (*TN*) are made up of a set of finite automata and represented within a graph system. The edges indicate transitions and the nodes the states of the single automata. Each automaton stands for a non-terminal symbol and is represented by its own network. The edges of each single network are denoted by non-terminal or terminal symbols and thus refer to other networks or final states. If the structure of a transition network also allows for recursive processes, for example, in the substitution of an object by another object belonging to a higher hierarchy level (e.g. a verb becomes a verbal phrase), this type of network is known as a *recursive transition network*. A path traversing the transition network starts at a first network and, beginning at the starting node, passes along the single edges. When it encounters a non-terminal symbol, the system branches like a sub-program to the corresponding network until finally all non-terminal symbols have been substituted. If different substitution possibilities are available, several paths between starting state and final state of the respective finite automaton exist. Figure 5.1 shows a transition network for expressions in natural language which may generate expressions such as "conductor likes singer," "a singer hates the conductor," "a singer likes a conductor hates the singer".

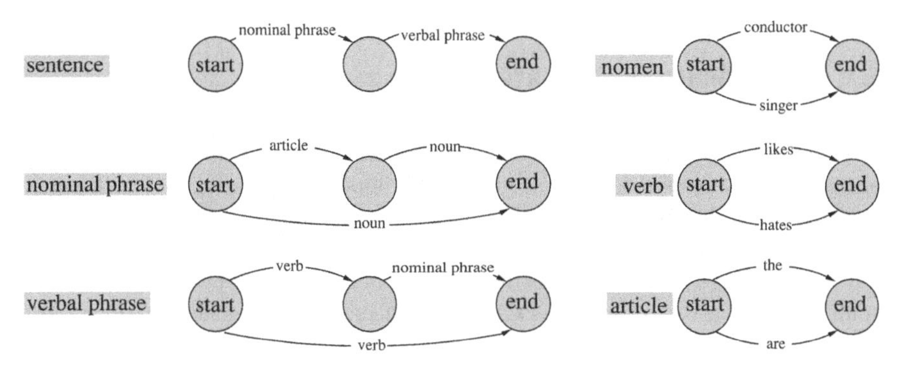

Fig. 5.1 A transition network for natural language expressions.

In an *augmented transition network* (*ATN*), the TN is extended in a way that allows specific instructions, conditional jumps or also whole sub-programs to be assigned to the edges. Augmented transition networks which were developed in the 1960s [11], are equivalent to type-0 grammars in terms of their generative capacity. Figure 5.2 illustrates an example of a simple ATN for the generation of melodic phrases. As an additional condition, the command "jump" is introduced here, inducing the omission of the current node. The upper and lower half of the graph enable generations in differing meters; a possible bass accompaniment is indicated below by means of some segments.

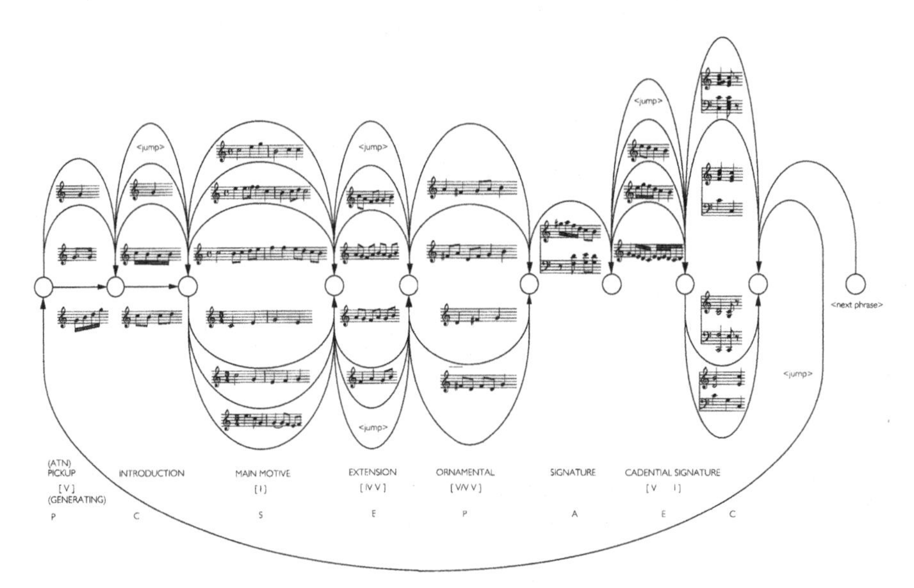

Fig. 5.2 Simple ATN for the production of musical segments [2, p. 64–65]. Reproduced with kind permission by A-R editions.

5.1 Experiments in Musical Intelligence

David Cope's Experiments in Musical Intelligence (EMI) is a well-known system of algorithmic composition which generates compositions conforming to a given musical style. Since its creation in 1981, Cope has continuously advanced EMI and described it in great detail in a number of publications.[1] EMI joins a series of different approaches to musical structure genesis and is also often mentioned in the context of artificial intelligence – Cope himself presents his system amongst oth-

[1] E.g. articles [1], [3]; book publications [2], [4], [5], [6].

ers in the framework of a "musical Turing test."[2] In view of the efficiency of EMI, Douglas Hofstadter revised some of his assumptions regarding the musical-creative potential of computer programs.[3] First experiments with a rule-based system for the generation of four-part movements Cope found unsatisfactory; as a consequence, he developed the approach of musical "recombinancy": In analogy to the historical model of the musical dice game, musical components are arranged to form a new composition, but with the essential difference being that EMI detects the components autonomously by means of the complex analysis of a corpus, transforms them partly and recombines them in an extensive process. For Cope, this principle also represents the implementation of a personal musical credo: "This program thus parallels what I believe takes place at some level in composers minds, whether consciously or subconsciously. The genius of great composers, I believe, lies not in inventing previously unimagined music but in their ability to effectively reorder and refine what already exists." [4, p. 93ff]. Since Cope implements the complex strategies of recombination within an augmented transition network, descriptions of basic functions of EMI are covered in this chapter.

Fig. 5.3 David Cope and Douglas Hofstadter. With kind permission of David Cope and Douglas Hofstadter.

In initial experiments, Cope divides up Bach chorales into single harmonic segments and recombines them by considering the correct voice leading, using only the transitions of harmonic segments which also occur in the original chorales. To provide sufficient material for new generations, the chorales of the corpus are transposed to a single key before analysis; the key for new compositions is chosen with regard to occurring voice ranges and the like. Although this simple principle of combination creates correct chorale progressions in either case, an acceptable structuring of a whole composition cannot be achieved. Due to this, additional strategies are applied that treat the musical material under numerous aspects for analysis and generation. In order to obtain a universally acceptable structure, a chorale of the

[2] See chapter 10; for Cope's test, cf. section "The Game" in [6, p. 33ff].

[3] Hofstadter refers here to his own prognoses in his book "Gödel, Escher, Bach: an Eternal Golden Braid" [8], cf. Hofstadter's essay "Staring Emmy Straight in the Eye ...," in [6, p. 33ff].

corpus may serve as a model for the sequence of phrases, cadence progressions, and the like. This model of chorale, however, only represents a meta-structure which is assigned with concrete musical segments from the EMI database. This database contains the complete material of the corpus, divided up into musical segments of different meaning. For the extension of the database and consequently also to increase the generalization power of EMI, coherent musical variations may be generated from segments of the corpus.[4] In order to be able to analyze, represent and process musical information in terms of different aspects, the following components and strategies are applied: SPEAC, an analysis model that analyzes the formal meaning of musical segments of different length on different hierarchical levels and makes them accessible for resynthesis; recognition and indication of characteristic and form-determining movements for their application in original or modified form in the new generations; implementation of the recombination strategies within an augmented transition network.

The musical units are recombined to form compositions in the relevant style in accordance with syntactic and "semantic" criteria. Here, the syntax describes allowed combinations of the terminals, whereas "semantics" guarantee that these also fulfil reasonable formal functions on their positions. Syntactic correctness in the combination of musical units can be seen through melodic components, for example in the transitions between the single parts. If, for example, a particular phrase in the corpus passes to another phrase through a major second movement, then this voice leading will also be retained in the recombination. This means that, in this case, only melodic segments are used as successors that may be reached from the last note of the previous phrase through a major second step. The syntactic correctness of a structure generated this way may be compared with the well-formedness of an expression generated by a generative grammar. Irrespective of that, the problem of "semantically" coherent meaning must be solved. If a generated expression is musically meaningful in the sense of musical semantics, it is defined in Cope's work by a sequence of musical components that due to their respective positions fulfil coherent forming and structuring functions. In functional harmony, for example, cadences that are characterized by dominant-tonic progressions would follow the "semantic" scheme of tension and relaxation. The imprecise terms "tension" and "relaxation" are used here according to Cope's classification system, which uses a general terminology to allow description of musical aspects on different formal levels. The semantic classification of the material is carried out by a system called SPEAC which is an acronym of the terms "statement," "preparation," "extension," "antecedents" and "consequent." These elements are used to indicate the musical units and denote relations that these may have to each other. "Statements" represent units that do not exist as the result of a particular process, but are included in a context in the course of the processing. "Preparations" are introductory gestures that stand ahead of other components and modify their meaning. "Extensions" follow other units (other than "preparations" or "extensions") and extend the preceding musical material. "Antecedents" prepare a concrete musical situation and demand

[4] Such as e.g. diatonic transposition which may represent a musical segment in the framework of different interval constellations; cf. [6, p. 102ff].

"consequents" as a solution. Figure 5.4 shows the different categories of SPEAC by means of a harmonic progression.

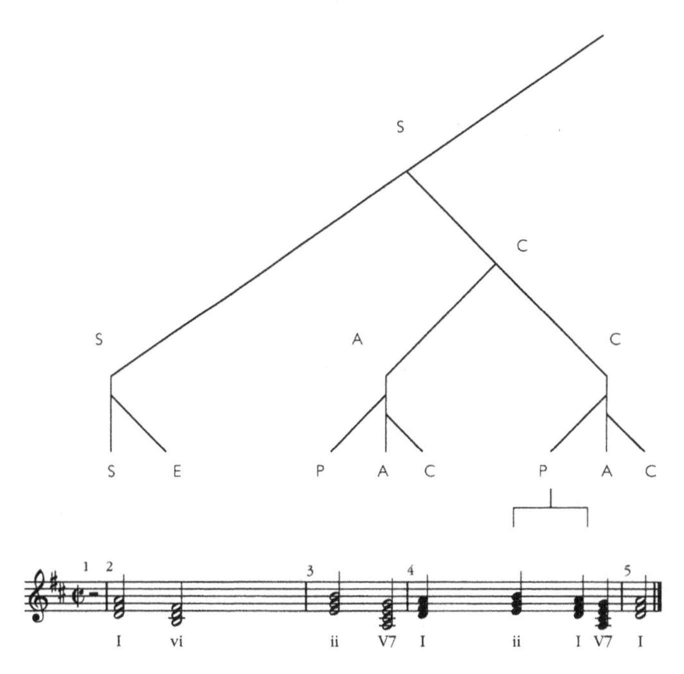

Fig. 5.4 Example of a harmonic parsing by SPEAC [2, p. 34]. Reproduced with kind permission by A-R editions.

The tonic parallel (h minor) is, for example, considered as the extension of the tonic due to its third relationship with D major. In this structure, the dominant seventh chord A7 in measure three is the "preparation" and the tonic "consequent." Here, the abstractions of SPEAC are assigned to the musical units on the basis of the intervallic constellation, the metric position and the duration of the single events. The order of the components of SPEAC is restricted by the following rules that indicate the possible successors of a component:

S → P, E, A
A → E, C
P → S, A, C
C → S, P, E, A
E → S, P, A, C

The SPEAC system is inspired by the analysis methods developed by Heinrich Schenker (see chapter 4) and allows for the interpretation of musical segments on different hierarchical levels. The dominant seventh chord in figure 5.4, for example, represents in the context of the two adjacent chords the function "antecedent" after "preparation" and ahead of "consequent." On a higher hierarchical level, these three chords represent an "antecedent" after a "statement" (measure 2) and before the final cadence which acts as a "consequent." This segmentation continuously involves increasing context dependencies in the examination, until finally a complete composition of the corpus is indicated as a first "statement," similar to the starting symbol "S" in a generative grammar. If during recombination a musical terminal[5] should be placed at a particular position, it is selected from the corpus according to syntactic conditions as well as the semantic context. In case, for example, the musical terminal should fulfil the function "PASC"[6] at this specific position, it is precisely these terminals chosen from the corpus that are designated with the classification "PASC" by EMI. If this selection turns out to be unsatisfactory, terminals with same properties but on a lower formal hierarchical level are searched, namely musical segments that fulfil the function "PAS," etc. If the formal context of the musical units in the corpus corresponds to that in the recombination, it must also be considered that, in order to achieve innovative solutions, the order of the terminals selected for new generations does not correspond to longer segments of the corpus.

To increase style conformity, EMI, during the analysis and recombination, also processes musical material which is of specific structuring meaning. "Signatures," "earmarks" and "unifications" are constellations that are prominent characteristics of a particular musical style. Furthermore, they may indicate distinctive changes in the musical sequence or serve for the internal structuring of a composition. "Signatures" are musical phrases that usually consist of characteristic melodic, harmonic and rhythmic components and often occur several times in a composition, usually in a modified form. EMI reveals "signatures" in the corpus using variably configurable pattern-matching processes, transforms them, if necessary, and also applies them in the process of generation in suitable positions within the composition. Since "signatures" may occur in different concrete musical shapes, EMI allows for their recognition and treatment within certain tolerances, meaning that, for example, a particular musical constellation on different scale steps or also in different rhythmic form may still be recognized as a consistent motif. A way to control the parameters of the pattern matching algorithms is given by the search for a particular number of "signatures" as it is typical of a composition of the respective style. The tolerance limits for identifying a signature are in this case extended until the desired number of "signatures" has been found. Furthermore, EMI allows for a corresponding manipulation of existing "signatures" of the corpus in the form of voice exchanges, different possibilities of pitch transposition, rhythmic refiguring and the like during the process of generation. "Earmarks" are characteristic movements that indicate the end or the beginning of a new formal segment of a composition. The consistent use of "earmarks" in recombination allows for a style-compliant segmentation of the material by means of musical signals such as particular cadence movements or trills. "Unifications," finally, are musical configurations whose structure is important for the internal structuring of a composition and therefore relate only to formal

[5] Meaning an expression which cannot be substituted anymore, cf. chapter 4.

[6] For "preparation," "antecedent," "statement," "consequent" as a path in the direction of the root of the graph which shows the formal functions of the musical terminal on different hierarchical levels.

elements of a single work. These patterns enable, for example, the favorable placing of a significant formal segment.

These characteristic movements are treated separately by EMI in the recombination process in order to maintain their structural integrity. Because different structural variants of, for example, motifs are recognized by EMI by means of pattern matching algorithms, they may for the application in recombination also be subjected to adequate transformations: "As the second movement started, I heard a very striking chromatically descending eight-note motive in midrange, then moments later heard the same motive way up high on the keyboard, then once again a few notes lower, [...] These widely spread entries gave an amazing feeling of coherence to the music [...] Astonished, I asked Dave what was going on and he replied, 'Well, somewhere in one of the input movements on which this movement is drawing, there must be some motive – totally different from this motive, of course! – that occurs four times in rapid succession with exactly these same timing displacements and pitch displacements' [...]".[7] Figure 5.5 shows shows the beginning of a fugue generated by EMI based on the corpus of the fugues from J. S. Bach's "Well-Tempered Clavier."

5.2 Petri Nets

Petri nets[8] are a special type of transition network that is used for the simulation of event-controlled processes and are represented by bipartite graphs. Nodes may consist of data, conditions and states (*places*) or actions (*transitions*). Transitions process data from places and store it in new places. The structure of the net results from the *flow relation* which relates particular places (generally represented by circles) and transitions (generally represented by rectangles) to each other by means of directed edges. The current state of the system is indicated by *tokens* that are distributed at specific places. After the net has been initialized through the *marking* of particular places, transitions may start to act by a process referred to as *firing*. When a transition fires, it takes the tokens from its input places and puts them on its output places; in other words, information is taken from places, processed and placed at other places. Figure 5.6 (top) shows the scheme of a calculation process by means of a simple Petri net and the chronological order of the markings (figure 5.6, bottom).

---

[7] Douglas Hofstadter in a conversation with David Cope; in [6, p. 50].

[8] In an originally simple form also called *condition nets* or *event nets*; developed as a mathematical representation of distributed systems in the 1960s by Carl Adam Petri and first introduced in his doctoral thesis; cf. [10].

Fig. 5.5 Beginning of a fugue generated by EMI. Example kindly provided by David Cope.

5.2.1 Petri Nets in Algorithmic Composition

Goffredo Haus and Alberto Sametti [7] developed ScoreSynth, a system of algorithmic composition that enables the processing of musical information with Petri nets. By means of interconnecting "music objects" (the places) with some transforming functions (the transitions), ScoreSynth can generate and manipulate control data in the form of MIDI values in different ways. The "music objects" consist of sequences of notes with associated information on pitch, duration, velocity and MIDI channel; the transitions enable manipulation of them by crescendo, decrescendo, crab, different possibilities of transition and the like. Because in a traditional Petri net a temporal structuring of sequences is not encoded (since the transitions fire in the moment they are connected with a marked place on the input side), the places are equipped with a counter which enables access to the information of the respective "music object" only after a certain period of time. For the programming of ScoreSynth, a special syntax is developed; furthermore, the possible application possibilities of

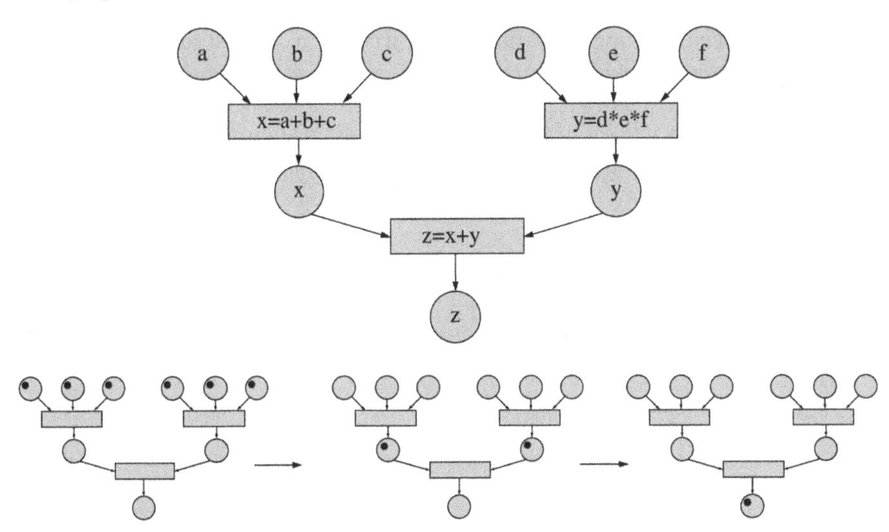

Fig. 5.6 Simple Petri net for arithmetic operations.

the software are extended by considering subnets and recursive net connections that may be restricted by a parameter controlling the number of recursions.

An interesting application of Petri nets in algorithmic composition is described by Douglas Lyon [9] who uses this formalism for modeling Markov chains of different order. The transition matrix of a Markov model is represented in a Petri net by different probabilities of weighted edges – an advantage of this approach shows in the fact that in the Petri net only transition probabilities $p_n \neq 0$ need to be processed [9, p. 19ff].

5.3 Synopsis

For the advantages and disadvantages of transition networks regarding tasks of algorithmic composition, in general the same principles apply as for generative grammars. The augmented transition network, for example, equals the type-0 grammar in terms of its expressive power. In contrast to genetic algorithms or cellular automata, for example, whose strong points become apparent in the realization of very specific compositional concepts, transition networks may be used in a broad field of musical structure genesis. It is exactly this aspect of the "universal" applicability of systems like these that makes them especially suitable for the development of algorithmic composition systems which enable the formulation of different compositional strategies in the sense of a programming language or the realization of a complex system design, as in EMI. An essential difference between a TN and generative grammars, and also Lindenmayer systems, may be seen in the representation of musical information within a graph. This difference also shows in the design of the user interface of computer music systems where– in contrast to a compact formulation of instructions within a text-based system – visual objects are manipulated. The graph representation of transition networks finds its parallels in computer music systems such as MAX,[9] PureData[10] or OpenMusic[11] that enable the manipulation of the musical information within a graphically interconnected structure.

References

1. Cope D (1987) An expert system for computer-assisted composition. Computer Music Journal 11/4
2. Cope D (1991) Computers and musical style. Oxford University Press, Oxford. ISBN 0-19-816274-X
3. Cope D (1992) On algorithmic representation of musical style. In: Balaban M, Ebcioglu K, Laske O (eds) (1992) Understanding music with AI. AAAI Press/MIT, Cambridge, Mass. ISBN 0262-52170-9
4. Cope D (1996) Experiments in musical intelligence. A-R Editions, Madison, Wis. ISBN 0-89579-314-8
5. Cope D (2000) The algorithmic composer. AR-Editions, Madison, Wis. ISBN 0-89579-454-3
6. Cope D (2001) Virtual music: computer synthesis of musical style. MIT Press, Cambridge, Mass. ISBN 0-262-03283-X
7. Haus G, Sametti A (1991) SCORESYNTH: a system for the synthesis of music scores based on Petri nets and a music algebra. IEEE Computer, 24/7, July, 1991
8. Hofstadter D (1979) Goedel, Escher, Bach: an eternal golden braid. Basic Books, New York. ISBN 0465026850
9. Lyon D (1995) Using stochastic Petri nets for realtime nth-order stochastic composition. In: Computer Music Journal 19/4, 1995
10. Petri CA (1962) Kommunikation mit Automaten. Schriften des Instituts für angewandte Mathematik, Bonn
11. Woods WA (1970) Transition network grammars for natural language analysis. Communications of the ACM, October 1970

[9] See: http://www.cycling74.com/.

[10] See: http://puredata.info/.

[11] A software specialized in algorithmic composition, developed by IRCAM: http://ircam.fr/; also see chapter 10.

Chapter 6
Chaos and Self-Similarity

Chaos theory became extremely popular in the 1980s due to a wide adoption of some aspects in the works of Edward N. Lorenz[1] and Benoit Mandelbrot[2] – the so-called "butterfly effect," self-similarity or the graphically fascinating illustrations of different fractals became the subjects of a broad non-scientific discussion as well. Regardless of whether the shape of coastlines, the branches of blood vessels or the complex behavior of dynamic systems are represented, chaos theory is occasionally given the significance of a "deus ex machina" – a universal explanation model for complex "natural" structures and processes. The euphoria for this discipline is also reflected by the title of James Gleick's book "Chaos: Making a New Science," [6] where the author predicted a paradigm shift in physics evoked by chaos theory.

Essential parts of chaos theory include the behavior of complex systems, their attractors as well as different forms of self-similar structures, above all of fractals. Self-similar structures may very well be modeled by Lindenmayer systems (LS). These were originally developed as a formal language for the description of the growth process of plants and present a powerful tool for the generation of musical structure in the field of algorithmic composition.

6.1 Chaos Theory

The term "chaos" derives from Greek and originally meant "space" or "abyss"; today, the word is colloquially used in the sense of "disorder," a meaning the term obtained in the course of the 17th century. In a mathematical and physical context, particular states of a system that are difficult to predict are called "chaotic." Chaos theory in a narrower sense is also referred to as theory of non-linear dynamics. In 1975, the term "chaos" was introduced in the field of mathematics through a publication of the mathematicians Tien-Yien Li and James Yorke [26].

[1] American meteorologist (1917–2008).

[2] French mathematician of Polish origin, born 1924.

The fact that even small modifications on the initial conditions may cause an unforeseeable behavior of a system was already shown by Jules Henri Poincaré[3] with his works on celestial mechanics.

Fig. 6.1 Jules Henri Poincaré. akg-images.

In his work "Science et méthode," Poincaré writes: "A very small cause which escapes our notice determines a considerable effect that we cannot fail to see, and then we say that the effect is due to chance. If we knew exactly the laws of nature and the situation of the universe at the initial moment, we could predict exactly the situation of that same universe at a succeeding moment. But even if it were the case that the natural laws had no longer any secret for us, we could still only know the initial situation approximately. If that enabled us to predict the situation with the same approximation, that is all we require, and we should say that the phenomenon had been predicted, that is governed by laws. But it is not always so; it may happen that small differences in the initial conditions produce very great ones in the final phenomena." [18, p. 67–68].

Another example for unpredictable chaotic behavior of a system can already be found in 1837 in Pierre-François Verhulst's[4] *logistic equation*. Verhulst introduced this equation as a demographic model which represents the temporal development of a population under the influence of different determining factors. If sufficient food is available, a particular population grows to a size at which food resources run short and a part of the population dies of starvation. For the diminished population, there is now again enough food available and it starts to increase again. Verhulst denotes this recurrent cycle within an equation representing the population size x at time $t+1$ in dependence on the population size x at time t. The threshold value for the limitation of x by the food supply is set at 1, and r is a constant – a product from an intrinsic growth rate and a value which presents a measure for the decimation of the population through starvation. The following equation results:

[3] French mathematician, physicist, and philosopher (1854–1912).

[4] Belgian mathematician (1804–1849).

$$x_n + 1 = rx_n(1 - x_n).$$

This equation represents a simple model of an ecosystem whose development is determined by a recurrent process. If the population is low, the factor $(1 - x_n)$ lies near 1, which enables an almost exponential growth. Accordingly, with a larger population, the factor $(1 - x_n)$ will approximate 0 – consequently, the population starts to decrease. The behavior of this system may be illustrated by means of a *Feigenbaum diagram*[5] (figure 6.2) showing the different limiting values of the population sizes dependent on r. Starting at a value of $r > 3$, duplications of accumulation points are increasingly produced, until finally with values for r between 3.57 and 4, chaotic behavior begins.

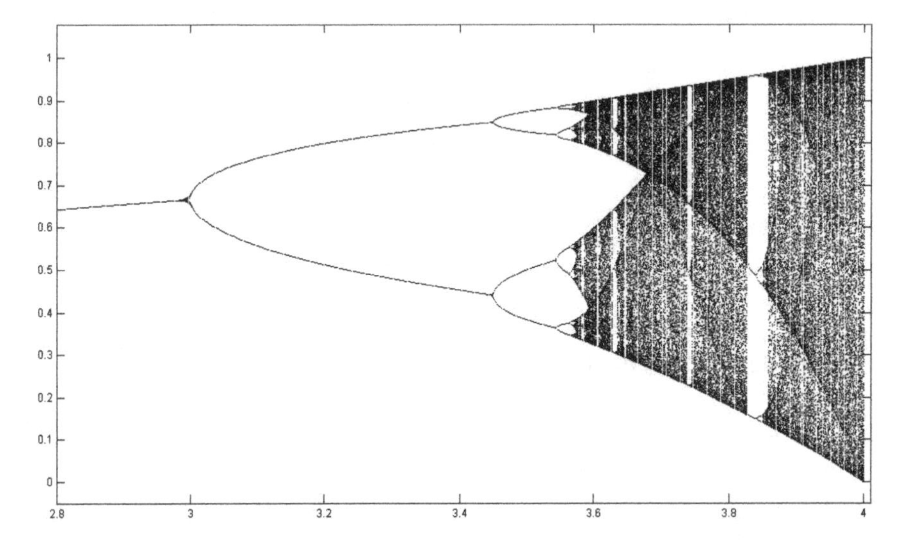

Fig. 6.2 Feigenbaum diagram of the logistic equation.

The best-known example of complex system behavior goes back to Lorenz who in 1963 developed a system of three coupled non-linear differential equations as a simplified model for atmospheric flow. Small changes performed on the values of the variables lead to completely different results, i.e. temporal developments of the system. This high sensitivity of these so-called *deterministic chaotic systems* in regard to smallest modifications in the initial conditions, are illustrated by Lorenz with the "butterfly effect": One flap of the wings of a butterfly causes a minimal turbulence which, however, in the course of the deterministic chaotic development of the system, may lead to completely unforeseeable meteorological consequences also in very distant places.

[5] Named after the American physicist Mitchell Jay Feigenbaum (born 1948).

Fig. 6.3 Edward Lorenz. With kind permission of Edward Lorenz.

6.2 Strange Attractors

When observing the long-term behavior of dynamical systems, the states of the system approach particular possible solutions. In other words, the *phase space* of the system evolves to a comparatively small region, which is indicated by the *attractor*. Geometrically, simple attractors may be *fixed points*, such as in a pendulum, for example, which evolves towards its resting state in the lowest point of the track. Another form would be the *limit cycle* in which the solution space is a sequence of values that are run through periodically. These simple attractors have in common that they have an integer *dimension* (see below) in the phase space. The structure of so-called *strange attractors* reflects the behavior of chaotic systems – they cannot be described with a closed geometrical form and therefore, since they have a non-integer dimension, are *fractals* (see below). Well-known examples of strange attractors as a representation for the limiting values of non-linear equation systems are the Hénon attractor, the Rössler attractor and the Lorenz attractor (figure 6.4), whose form resembles a butterfly.

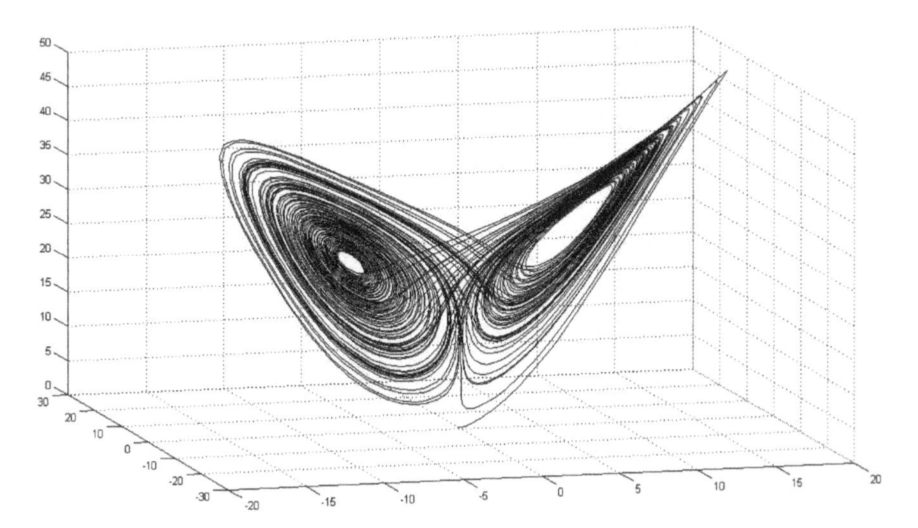

Fig. 6.4 Lorenz attractor.

6.3 Fractals

Fractals are geometric shapes that show a high degree of self-similarity (also: *scale invariance*), meaning that particular graphic patterns reoccur in identical or very similar shapes on several different orders of magnitude. Fractal structures can be found in processes such as crystallization, the shape of coastlines or also in numerous manifestations of plant growth, e.g. in the form of ferns or particular variants of cauliflower. An interesting treatment of self-similar structures in the field of painting can be found in the work of the Dutch painter Maurits Cornelis Escher (1889–1972). The term "fractal" was introduced by Benoit Mandelbrot: "I coined *fractal* from the Latin adjective *fractus*. The corresponding Latin verb *frangere* means to "break": to create irregular fragments. It is therefore sensible – and how appropriate for our needs! – that, in addition to "fragmented" (as in *fraction* or *refraction*), *fractus* should also mean irregular, both meanings being preserved in *fragment*." [14, p. 4].

In the 1980s, it was above all the fascinating graphic representations of fractals, especially of the Mandelbrot set, that aroused strong interest in these structures. Self-similar shapes in the field of mathematics were, however, already developed a long time before Mandelbrot. The *Cantor set*[6] (figure 6.6) is a closed subset of real numbers and is in its graphic representation a fractal with a simple generation instruction: Delete the middle third from a set of line segments and perform this iteration step on each resulting new line segment.

[6] Introduced by and named after the German mathematician Georg Cantor.

Fig. 6.5 Benoit Mandelbrot. With kind permission of Benoit Mandelbrot.

Fig. 6.6 Iteration steps of a Cantor set.

Another example of a fractal is the *snowflake curve* which was developed by the Swedish mathematician Helge von Koch (1870–1924) already in 1906.[7]

A common aspect of all fractals is their broken *dimension*.[8] In mathematics, the term dimension refers in general to the degrees of freedom of a movement in a space. Accordingly, a line is one-dimensional, whereas an area has two dimensions. A fractal line pattern, however, may in the course of the iteration process become more and more like a plane; therefore, the fractal dimension of this shape lies between 1 and 2.

Regardless of the variety of fractal shapes, it is nevertheless the *Mandelbrot set* which enfolds an extreme complexity in its graphic representation and due to a number of aesthetically appealing realizations is considered the "prototype" of a fractal. Mandelbrot developed with the set named after him, which often is also referred to as "Apple Man," a possibility to classify *Julia sets*[9] that represent subsets of the complex numbers. The iterative construction rule[10] of this set with the initial condition $z_0 = 0$ and representing c as a complex number, is as follows:

[7] See section "Lindenmayer Systems."

[8] For a detailed description of *fractal dimension*, see [14, p. 14ff].

[9] Named after the French mathematician Gaston Maurice Julia (1893–1978).

[10] When a system consists of several iterative equations, it is also referred to as *iterated function system* or *IFS*.

$$zn + 1 = zn^2 + c.$$

The Mandelbrot set is a dynamic calculation based on the iteration of complex numbers. For each position or number, a value for a particular amount of iterations is determined. All positions, whose values remain finite also after a large number of iterations, make up the Mandelbrot set and may, for example, be pictured by black points. Depending on the number of iterations that yield a value for zn which is higher than a defined limiting value, the respective positions may also be represented by different colors or shades of grey. According to the segment of the number level, sections of the Mandelbrot set may be "scaled up," this "zooming" (figure 6.7) making continuously new self-similar structures visible.[11]

Fig. 6.7 A zoom into the Mandelbrot-set.

6.4 Lindenmayer Systems

Lindenmayer systems (*LS*) or *L-systems* are named after the botanist Aristid Lindenmayer (1925–1989) who developed a formal language to represent the growth

[11] Very appealing graphic representations of Mandelbrot sets and other fractals can be found in [17].

Fig. 6.8 Aristid Lindenmayer [21].

of algae in 1968 [11]. In 1974, Paulien Hogeweg and Ben Hesper extended this system by introducing a graphic representation system [8]. Beginning in 1984, L-systems have been used by Alvy Ray Smith for representing the growth processes of plant structures [23]. Other works in the field of growth simulation with L-systems were produced by the mathematician Grzegorz Rozenberg and the computer scientist Przemyslaw Prusinkiewicz.[12]

Similar to the grammars of the Chomsky hierarchy, L-systems (also: *Parallel Rewriting Systems*) work with rewriting rules. Based on a starting element, production rules are applied whose output is usually represented graphically. In contrast to the implementation of grammars, L-systems do not distinguish between terminal and non-terminal symbols; furthermore, all production rules and rewriting rules are applied simultaneously. In a Lindenmayer system symbols are for the most part replaced by symbol strings that in turn again contain these symbols. As a consequence, in comparison to a generative grammar, the number of symbols to be processed increases enormously. Due to these basic characteristics, Lindenmayer systems are also well suited to model self-similar structures. and may most simply be represented as a triplet (v, ω, P), see table 6.1.

The application of a derivation is based on an axiom. The production rules are applied by substituting the single preceding symbols with their successors. If a particular predecessor does not have a successor, then $\alpha \rightarrow \alpha$ applies, meaning that the predecessor is replaced by itself.

[12] A good introduction in the theory of Lindenmayer systems can be found in the work of Prusinkiewicz and Lindenmayer "The Algorithmic Beauty of Plants" [21].

v	the alphabet: a finite set of symbols, most commonly noted in lower-case letters, such as: $v = \{a,b,c,d\}$
v^*	the set of all possible symbol strings from V, such as: *aabc aab aba acccc*, etc.
v^+	the set of all possible symbol strings without the empty set $v^* \backslash \{\emptyset\}$
ω	the axiom or the *initiator*; it is applied where $\omega \in v^+$
P	a finite set of production rules or *generators*
	The production rules are mostly represented by $\alpha \rightarrow \chi$ with a predecessor $\alpha \in v$ and a successor $\chi \in v^*$

Table 6.1 Definitions for a Lindenmayer system.

6.4.1 Forms of Lindenmayer Systems

Some basic types of L-systems can be distinguished:

- *Context-free* and *context-sensitive L-systems*
- *Deterministic* and *stochastic L-systems*
- *Parametric* and *non-parametric L- systems*

In order to graphically represent[13] L-systems, symbols that mark out the motions of a *turtle* by using line graphics are for the most part used. This visual representation style is called *turtle graphics*, originally written in the programming language LOGO. In its simplest form, the turtle has the following commands (table 6.2):

F	Move forward one step. A line segment is drawn
φ	Angle for the turning of the turtle
$+$	Turn left by angle φ
$-$	Turn right by angle φ
[	Save current values for position and angle
]	Restore the saved values
I	Iteration depth

Table 6.2 Turtle commands.

[13] All graphics in the following examples have been designed with the program "Virtual Laboratory," [22] which in contrast to most other programming environments also enables the generation of stochastic and parametric L-systems. The code for the various L-systems is taken in modified form from [15].

6.4.1.1 Context-Free and Context-Sensitive L-Systems

A Lindenmayer system is called a context-free L-system (also: *0L-system*) if the substitution is carried out independently of its environment. Together with a deterministic structure for the production rules, these simple L-systems form the class of deterministic 0L-systems (*D0L-systems*). In a context-sensitive L-system, the application of a particular χ depends on the environment of the α that is to be substituted. A possible notation of context-dependent production rules could be as follows:

$$P_1 = a > b \rightarrow aba$$
$$P_2 = a < b \rightarrow bab$$
$$P_3 = a < b > a \rightarrow baa$$

P_1: b is replaced by aba, when it is positioned before an a.
P_2: b is replaced by bab, when it is positioned after an a.
P_3: b is replaced by baa, when it is positioned before and after an a.

Often context-sensitive L-systems are also referred to as *IL-systems*. Within the class of IL-systems, 1L-systems consider one and 2L-systems two symbols that neighbor.[14] *(k, l)-systems* refer to production rules in which the left context of α consists of k, the right context of l symbols.

Figure 6.9 shows examples of a Koch curve[15] generated by a D0L-system, with different iteration depths and the following rules:

$$\varphi = 60°$$
$$axiom = F$$
$$P : F \rightarrow F+F--F+F$$

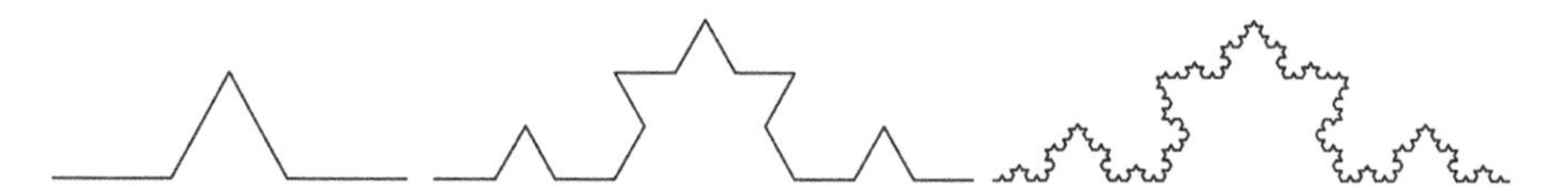

Fig. 6.9 Koch curves; recursion depths: 1, 2, 10.

If the set of rules is extended by a symbol f, which covers a distance without drawing a line, structures that are delimited from one another can be generated, as shown in figure 6.10.

$$\varphi = 90°$$
$$axiom : F+F+F+F$$

[14] Example: P1 and P2 for 1L, and P3 for 2L-systems.
[15] Developed by the Swedish mathematician Helge von Koch around 1906.

$P_1 : F \rightarrow F + f - FF + F + FF + Ff + FF - f + FF - F - FF - Ff - FFF$
$P_2 : f \rightarrow ffffff$

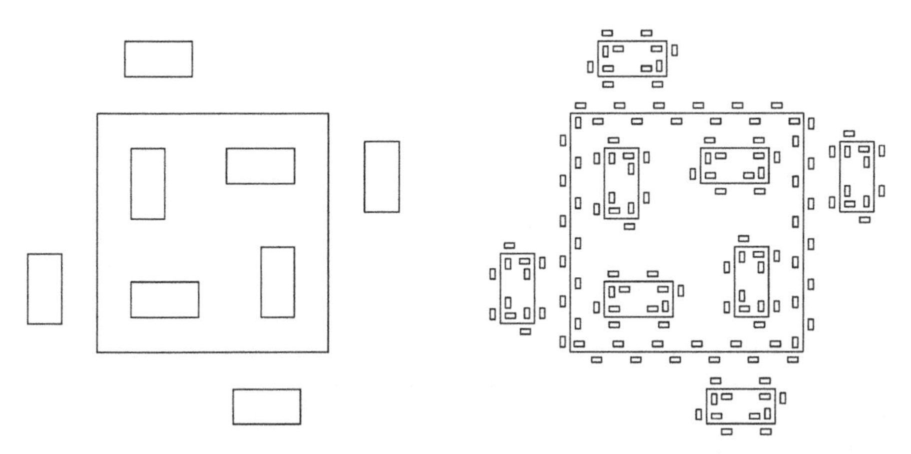

Fig. 6.10 Structure with recursion depths 1 and 2.

An example of a D0L-system is provided by the graphic representation of the dragon curve[16] which in this case can be generated by rewriting rules for two distances "F" und "G" of the same length and the following rules:

$\varphi = 45°$
$0\ axiom : F$
$P_1 : F \rightarrow -F + +G$
$P_2 : G \rightarrow F - -G+$

Due to the parallel substitutions, the derivations of *recursion depths* 1 to 3 in the dragon curve are (the brackets indicate the replaced expressions):

$RT_1 : -F + +G$
$RT_2 : -(-F + +G) + +(F - -G+)$
$RT_3 : -(-(-F + +G) + +(F - -G+)) + +((-F + +G) - -(F - -G+))$

Figure 6.11 shows the dragon curves with recursion depths 1, 2, 3, 5, 13.

[16] Also called Harter-Heighway Dragon, developed in the 1960s by the physicians John Heighway, Bruce Banks and William Harter.

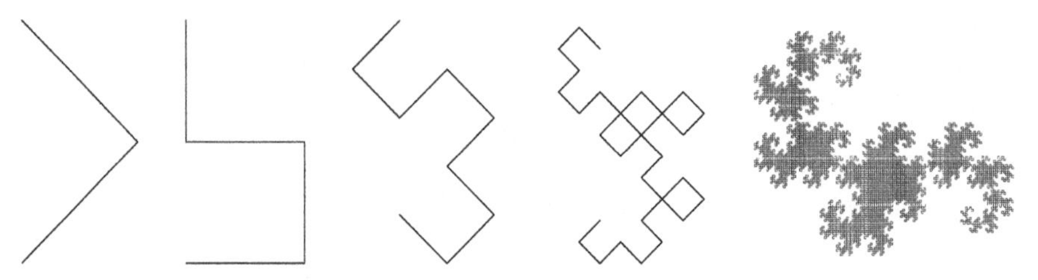

Fig. 6.11 Dragon curves with different recursion depths.

6.4.1.2 Deterministic and Stochastic L-Systems

In a *deterministic L-system*, there is only one $\chi \in V^*$ for each $\alpha \in V$. If χ may be chosen with a certain degree of freedom, the system is referred to as *stochastic L-system*, notated as quadruple $\{V, \omega, P, \pi\}$. The function π maps the corresponding predecessor (α) on probabilities of the successor (χ). The total of all probabilities of χ that are assigned to a particular α, must equal 1. The rewriting rules may be notated as follows:

$$P_1 = a \rightarrow^2 aba$$
$$P_2 = a \rightarrow^3 bab$$
$$P_3 = a \rightarrow^5 abb$$

Accordingly, a substitution of a is carried out for example with a probability of 20% by P_1, with a probability of 30% by P_2, and a probability of 50% by P_3.

Figure 6.12 illustrates an example of different structures of a stochastic L-system based on the following rules[17]:

$$\varphi = 45^\circ$$
$$axiom : F$$
$$P_1 : F \rightarrow F[+FF]F[-F]F : 1/3$$
$$P_2 : F \rightarrow F[+F]F : 1/3$$
$$P_3 : F \rightarrow F[-FF]F : 1/3$$

6.4.1.3 Parametric and Non-Parametric L-Systems

In a *parametric L-system* (also: *parameterized L-system*), the application of rewriting rules may be made subject to particular conditions. In this system, a production rule consists of a predecessor, a condition, and a successor. The symbols may change their values during the substitution processes, causing different production rules to

[17] The fractions after the colons indicate the probabilities for the selection of the production rule.

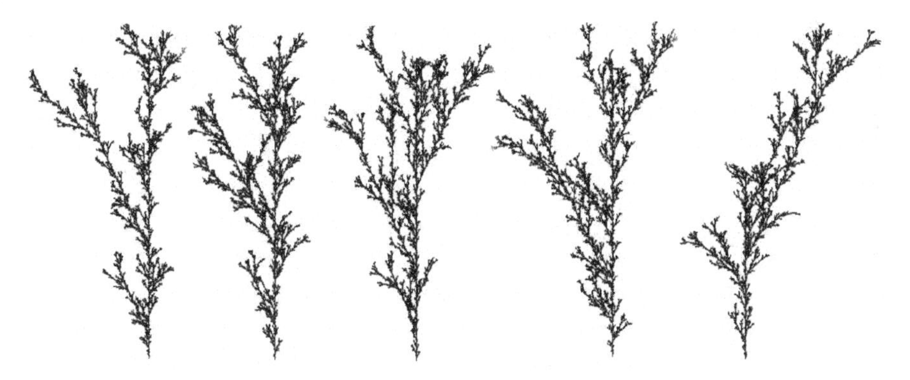

Fig. 6.12 Different generations of a stochastic L-system of recursion depth 7.

be chosen. The values that succeeding symbols assume can be made dependant on one another by a parametric L-system. In a musical context, these possibilities enable the formulation of rules such as, for example, "Transpose the preceding note one semitone" or "If the note is reached by a large interval step, only allow a second progression."

Figure 6.13 illustrates an example of a parametric L-system. Here, variables are indicated, a starting value for F is given and an abort condition is formulated.[18]

$$a = 86;\ p = 0.3;\ q = (1 - p);\ h = ((p * q)^{0.5});\ Axiom : F(0.8)$$
$$F(x) : x > 0.03 \rightarrow F(x * p) + (a)F(x * h) - (a + a)F(x * h) + (a)F(x * q)$$

Fig. 6.13 Example generation of a parametric L-system.

L-systems may also be applied to manipulate symbol strings so that the number of symbols does not necessarily need to increase in each substitution process.

[18] The syntax of "Virtual Laboratory" is represented here in a simplified form.

In this sense, for example, the following set of rules makes the letter a "travel through" a symbol string.

$Axiom: aXXXX$
$P_1 : a < X : X \rightarrow a$
$P_2 : a \rightarrow X$
(If X is positioned after an a, replace this X by a, replace a by X)

Results: $aXXXX$; $XaXXX$; $XXaXX$; $XXXaX$; $XXXXa$; XXXXX

6.5 Chaos and Self-Similarity in Algorithmic Composition

In the context of the heterogeneous field of chaos theory, different approaches are often assigned to algorithmic composition that deal with the musical realization of *fractional noise*, the mapping of fractals and attractors as well as different aspects of self-similarity. Not considering chaos theory in general, self-similar structures may in the field of algorithmic composition very well be modeled by Lindenmayer systems that, similarly to a generative grammar, enable the realization of complex compositional concepts.

6.5.1 Fractional Noise

A frequently applied form of musical structure generation from the field of chaos theory uses various shapes of what is known as fractional noise.[19] The term describes different forms of noise that are distinguished in regard to their *spectral density*, expressing the distribution of noise power with frequency. *White noise* (figure 6.14, top) is here characterized by the relation $1/f^0$ and describes a stochastic process of *uncorrelated* random values. As with a repeatedly thrown dice, the numbers on the dice sides are not associated with each other in any way. A highly *correlated* variant, meaning that the values in a sequence strongly depend on each other is *brownian noise* (figure 6.14, middle), showing a spectral density of $1/f^2$. Here, for example, only adjacent number values may succeed each other. The most interesting form in regard to musical structure genesis is *pink noise*, also referred to as *1/f noise* (figure 6.14, bottom), whose behavior lies somewhere between the abovementioned extremes.

[19] The term "fractional noise" was coined by Mandelbrot and the mathematician John W. Van Ness in 1968, cf. [12]. "Fractional noise" is often also referred to as "fractal noise."

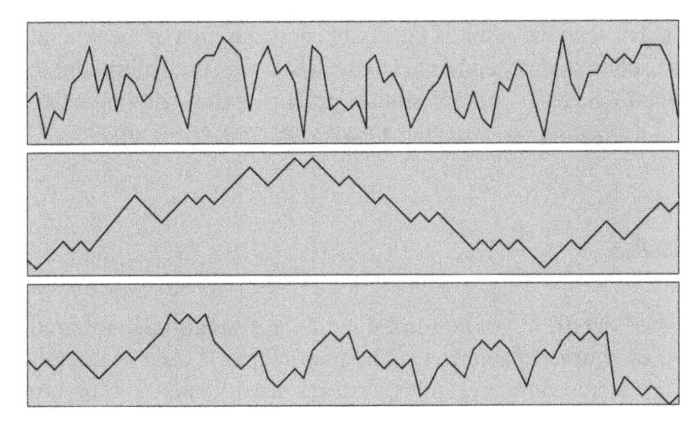

Fig. 6.14 The characteristics of white, brownian and pink noise.

In a musical mapping on, e.g. pitches, the characteristics of pink noise show as a progression in which stepwise movement and melodic jumps are in a well-balanced relation.

Richard F. Voss and John Clarke described characteristics of spectral density in recordings of different musical genres and showed their parallels to the peculiarities of $1/f$ noise: "The observations on music suggest that $1/f$ noise is a good choice for stochastic composition. Compositions in which the frequency and duration of each note were determined by $1/f$ noise sources sounded pleasing. Those generated by white noise sources sounded too random, while those generated by $1/f^2$ noise sounded too correlated." [25]. Voss and Clarke extended their 1-dimensional model to a two-voice structure, which is partly correlated and whose rhythmic shape may be also designed applying $1/f$ noise.[20] Based on the works of Voss and Clarke, Charles Dodge and Thomas A. Jerse [4, p. 368ff] described the generation of $1/f$ sequences and produced examples of musical mappings of these different noise forms.

Dodge [3] described a musical structure consisting of three melodic lines (voices) on the basis of $1/f$ noise. For each line, a particular number of different pitch classes is determined that must be produced by the output of the $1/f$ noise. This means that new pitch classes are produced as long as the desired number of different values has been reached. After the first line has been generated this way, the program creates a second line using the same procedure: For each note in the first line, a succession of notes is produced for the second line, until all notes of the second line have been generated. This process is also performed for the generation of the third line by producing a sequence of third line notes for every note in the second line. The result is a three-voice structure of increasing density. For determining the current durations, a fourth line is created whose notes, however, are not included in the score, but determine a rhythmic length for every note of the third line. If, for example, for the first note of the third line four notes of the fourth line would be assigned, and for

[20] For the musical implications of $1/f$ noise, also see [14, p. 374–375], for extensions of the approach of Voss and Clarke regarding possibilities of intervention, cf. [2].

the second note eight notes, then the result is the double note duration of the second note, etc. This process is consequently continued to the first voice, until all rhythmic values have been obtained. The result is a self-similar structure, whose rhythmic fine segmentation is created depending on achieving a particular "tonal diversity."[21]

6.5.2 Chaotic Systems

Jeff Pressing [19] mapped the *orbit*[22] of non-linear equation systems (also referred to as non-linear maps) on musical parameters. The map output is used to control pitch, duration, envelope attack time, dynamics, textural density and the time between notes of single events of synthesized sounds. So, for example, the population size (the output) of the logistic equation is converted to an appropriate pitch range through $F = 2^{(cx+d)}$, where the constant c equals the range in octaves and 2^d the lowest pitch produced (in hertz). Since the value of the population size lies in the range between 0 and 1, it is also directly used for the time between notes. Other parameters such as dynamics and envelope attack also yield from the resulting value of the equation by subjecting them to different arithmetic operations. In order to receive different but nevertheless correlated values for musical parameters, complex mapping strategies for equation systems with up to four dimensions are given. Figure 6.15 shows a musical structure resulting from a four-dimensional map.

Rick Bidlack [1] also mapped the orbit of 2-, 3- and 4-dimensional equation systems on musical parameters. The musical textures that, for example, may be obtained by means of a Hénon equation through the mapping of dimensions in phase space on musical parameters (e.g. pitch, duration, dynamics, etc.), however, for Bidlack do not present definitive musical results[23]: "Rather than viewing the output of chaotic systems as music in its own right, however, it is probably best to consider such output as raw material of a certain inherent and potentially useful musicality. Clearly there will be as many ways to apply chaos to musical decision as there are composers interested in doing so." [1, p. 2]. This interpretation of data of a structure-producing algorithm enables – not only in applications of chaos theory – creative approaches to the realization of individual compositional concepts.

Another interesting approach made by Jeremy Leach and John Fitch [10] derives a tree structure from the orbit of a chaotic system. The design of the tree structure being inspired by the works of Lerdahl and Jackendoff (see chapter 4), consists of a hierarchical arrangement of scales and note values and results from the interpretation of the values of the orbit as hierarchic positions of nodes, as illustrated in figure 6.16a. Concrete note values are produced by interpreting nodes of higher hierarchic

---

[21] An exhaustive approach regarding self-similar reductions of interval constellations can be found in a theoretic work by composer Bernhard Lang, cf. [9].

[22] Also called *trajectory*, the sequence of values that forms the result for a particular variable of an iterative equation system. These values mostly approach an attractor of a certain shape.

[23] In this context, also see the work of Michael Gogins, who interprets the representations of IFS as musical scores, cf. [7].

Fig. 6.15 Musical mapping of a 4-dimensional equation system by Pressing [19, p. 43]. © 1988 by the Massachusetts Institute of Technology.

order as pitches that structure a melodic progression. These pitches serve as turning points of the resulting melodic movement and also indicate, depending on their distance to the next segment (i.e. sub-tree), a different size for an interval segment (figure 6.16b).

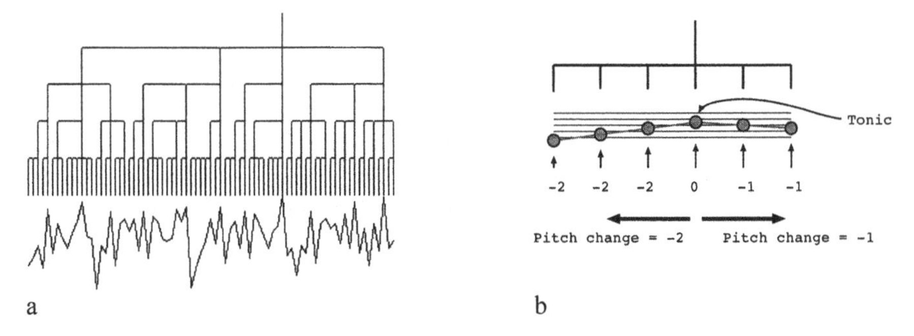

Fig. 6.16 Chaotic sequence with corresponding tree structure (a); melodic progression in a subtree (b) [10, p. 31, 29]. © 1995 by the Massachusetts Institute of Technology.

6.6 Lindenmayer Systems in Algorithmic Composition

In an early work, Przemyslaw Prusinkiewicz [20] described the simple mapping of the generation of note values from a turtle interpretation of Lindenmayer systems. Prusinkiewicz gives an example by means of a Hilbert curve. In the graphic, notes are represented by successive horizontal line segments; the lengths of the segments are interpreted as tone durations. The pitches result from the vertical position of the segments and are mapped on the steps of a C major scale (figure 6.17).[24]

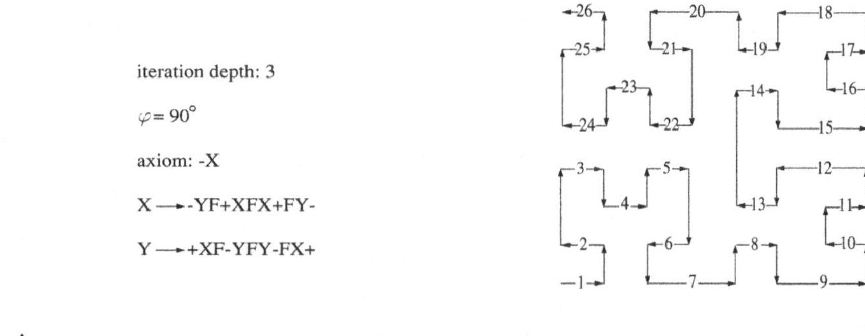

Fig. 6.17 Mapping of a Hilbert curve on scale tones.

John McCormack [16] compared stochastic approaches, Markov models, different variants of generative grammar, and Lindenmayer systems in terms of their suitability for musical production. In an extendable system design, McCormack introduces a program architecture for algorithmic composition. The system allows for the application of context-free and parametric L-systems; the involvement of hierarchically structured grammars allows for variable possibilities of musical structure genesis.

Hierarchical grammars are built up like D0L-systems; however, entire grammars may be used for single symbols of the successor. Although each of the different grammars expands independently, it is possible to establish structural relations between the single rewriting systems by means of parameterization. McCormack's system uses the notion of "virtual players," modules responsible for a voice or an instrument, and which are controlled each by their own rewriting system. The processing scheme is represented in figure 6.18: Reading and parsing, putting the current symbol string on the value of the axiom, recursive application of the substitutions on the current symbol string, and finally the output of the results as MIDI values.

In this system, the parameters tone pitch, duration, timbre, and different controllers may be controlled within a polyphonic structure. In order to do that, addi-

[24] This, however, actually only represents a musical mapping of a graphic interpretation of an L-system.

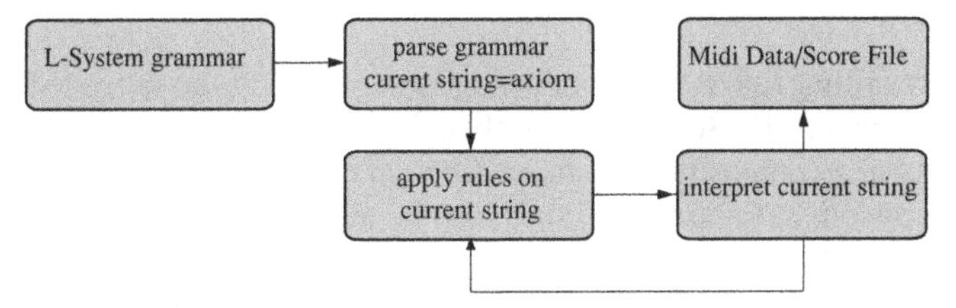

Fig. 6.18 Stages of processing in McCormack's system.

tional symbols are defined within the rewriting rules that, depending on the position, may express music-specific context dependencies. In this sense, for example, the expression "$(CE) \mid (GC) \rightarrow D(CE)$" means: If the current harmony consists of the notes G and C played simultaneously and they were preceded by C and E played simultaneously, then play D for the current duration followed by C and E played simultaneously.

6.6.1 Mapping Strategies of Different Lindenmayer Systems

In his dissertation, Roger Luke DuBois [5] described different possibilities of mapping Lindenmayer systems on musical parameters and consequently developed a real-time application in which a musician provides live data input. DuBois differentiates his approach from mapping strategies which are based on a realization of turtle graphics. In his system, turtle graphics serve only for visualizing the structure of the L-system; the L-systems are mapped on musical parameters by means of mapping symbols of an LS on musical units. An example for such a representation is shown by the following rewriting system in which a number of concrete musical events may be assigned to each single symbol of the LS.

$Axiom: X$
$Iteration\,depth: 5$
$P_1: X \rightarrow F - [[X] + cX] + F[+FcX] - X$
$P_2: F \rightarrow FF$

The following section shows a symbol string resulting in the fifth iteration step:

$FFFFFFFFFFFFFFFF - [[FFFFFFFF - [[FFFF - [[FF - [[F - [[X] + cX] + F[+FcX] - X] + cF - [[X] + cX] + F[+FcX] - X] + FF[+FFcF - [[X] + cX] + F[+FcX] - X] - F - [[X] + cX] + F[+FcX] - X] + cFF - [[F - [[X] + cX] + F[+FcX] - X] + cF - [[X] + cX] + F[+FcX] - X] + FF[+FFcF - [[X] + cX] + F[+FcX] - X] - F[[X] + cX] + F[+FcX]X] + FFFF[+FFFFcFF - [[F - [[X] + cX] + F[+FcX] - X] + cF - [[X] + cX] + F[+FcX] - X] + FF[+FFcF - [[X] + cX] + F[+FcX] - X] - F - [[X] + cX] + F[+FcX] - X] \ldots]$

The successor of rewriting rule p1 is prominently represented in the derivated symbol string; for this reason, DuBois applies a changing mapping which in different formal segments of a composition produces different characteristic pitch sequences, as shown in table 6.3. Figure 6.19 illustrates the musical mapping of the

Pitch Class	A	B	C	D	E	F	G	H	I
C0	F	[	[	-	(P)	F	c	[	
C#/Db1		]						]	
D2	]			+	+	]	[		+
D#/Eb3	(P)	F		]			(P)	F	
E4			-		-	(P)	F		-
F5	+	-	+		[	+		-	[
F#/Gb6		+					]	+	
G7	-		(P)	F	c	-	+		c
G#/Ab8		(P)	F	[					
A9	c		c	(P)	F	c	-	(P)	F
A#/Bb10	[	c	]	c	]	[		c	]
B11									

Table 6.3 Pitch mapping of Lindenmayer symbols.

string: $F-[[X]+cX]+F[+FcX]-X]+$; "X" is interpreted as a rest in the segments A and F (figure 6.19, top) as well as B and H (figure 6.19, bottom). Even though due to these different representations a consistent musical interpretation of the Lindenmayer system is abandoned, characteristic symbol strings may, however, be well recognized in different musical segments through similar melodic contours.

Fig. 6.19 Mapping on tone pitches and rests in different segments.

Another form of symbolic mapping for the generation of a polyphonic structure is illustrated by DuBois in the following example by means of a context-sensitive Lindenmayer system.

$Axiom : bab$
$P_1 : A < B > B \rightarrow BA$
$P_2 : A < A > B \rightarrow AB$
$P_3 : A < B > B \rightarrow B$
$P_4 : B < A > \rightarrow AA$

The following symbol strings result for the first substitutions:

$BAB(axiom)$
$BAAB$
$BAAABB$
$BAABABBAB$
$BAAABBAABABAAB$
$BAABABBABAAABBAABAAABB$

DuBois interprets each symbol as a successive semitone step over an underlying root tone; only for the letter *B* is a note set, *A* functions as a chromatic "gap," leaving out the respective note on that position (figure 6.20).

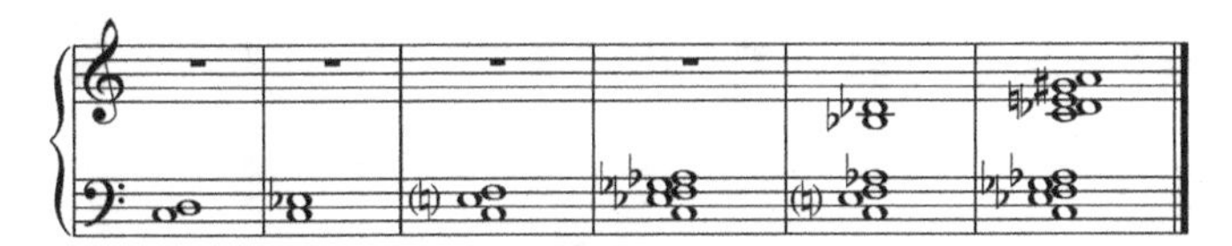

Fig. 6.20 Mapping for a polyphone structure.

The generated chords may be mapped melodically on a monophonic rhythmic structure (figure 6.21), while *B* determines a 1/16 note and *A* indicates a 1/16 hold, for example, *BA* from the axiom *BAB* becomes a 1/16 note with a 1/16 hold, as can be seen in the illustration, where the first note c gets the duration of 1/8.

Fig. 6.21 Mapping of chords on a monophonic melody.

For another form of musical representation, a context-sensitive L-system that produces symbol strings of the same length in all generations is applied:

$Axiom : 39 * W, B, 40 * W$
$P_1 : B < B > B \rightarrow W$
$P_2 : B < B > W \rightarrow W$
$P_3 : B < W > B \rightarrow W$
$P_4 : B < W > W \rightarrow B$
$P_5 : W < B > B \rightarrow W$
$P_6 : W < B > W \rightarrow W$
$P_7 : W < W > B \rightarrow B$
$P_8 : W < W > W \rightarrow W$

It is apparent that this structure may also be represented by a cellular automaton of the form:

states	111	110	101	100	011	010	001	000
following states	1	0	1	1	0	1	1	1

This is why an LS of this type may equal a 1-dimensional cellular automaton. These production rules generate the form of a "Sierpinski triangle."[25] In DuBois' system, the mapping is performed by the representation of the LS in a grid; tone pitches ascending from right to left in semitone steps are assigned to the cells of the x-axis. Figure 6.22 illustrates an extract of the musical result.

Fig. 6.22 Mapping of a "Sierpinski triangle" on polyphone structures.

For the mapping of parametric Lindenmayer systems, DuBois puts the succeeding substitutions in the context of the precedent musical events. The symbols of the rewriting rules are mapped relatively on the precedent event, so that, e.g. rules such as "Transpose the next note by a third" or "Decrease the current duration by a particular value" may be applied. The representation of pitches may be based on pitch classes or interval steps. The duration of each musical event is a relative value which may be proportionally shortened or lengthened by applying particular rules. This means that, for example, by having a current value of a quarter, after a particular symbol, all following events will have the duration of a dotted eighth note. Because, consequently, musical parameters such as pitch or duration will always be determined in dependence to the precedent events, same symbols of the LS will also create different musical results in the course of time.

DuBois indicates the following LS as an interesting example for the processing of tone pitches that in the course of the substitutions are reduced to some tonal centers:

$P1 : 0 \rightarrow 0$	$P4 : 3 \rightarrow 3$	$P7 : 6 \rightarrow 3$	$P10 : 9 \rightarrow 2$
$P2 : 1 \rightarrow 0$	$P5 : 4 \rightarrow 7$	$P8 : 7 \rightarrow 3$	$P11 : 10 \rightarrow 11$
$P3 : 2 \rightarrow 8$	$P6 : 5 \rightarrow 5$	$P9 : 8 \rightarrow 11$	$P12 : 11 \rightarrow 5$

[25] A triangle which is divided into four other triangles by means of recursive segmentation of a preceding triangle. This self-similar structure may also be generated by a 1-dimensional cellular automaton of rule 90, cf. chapter 8.

Applied to a chromatic scale, the number of tone pitches is reduced to five values after the third substitution process, as illustrated in figure 6.23.

Fig. 6.23 L-system producing tonal centers.

6.7 Synopsis

Structures that may be assigned to the field of chaos theory are in most cases represented musically by means of a mapping of trajectories of a phase space on musical parameters. Since the behavior of chaotic systems is difficult to foresee and reacts highly sensitive to changes made on the initial conditions, accordingly, the musical mapping may provide a wide range of results. A basic problem for the realization of individual compositional intentions by means of chaotic systems results from the lack of possibilities of intervention and structuring – as soon as the underlying equation system has been chosen and the initial values have been set, the system continuously produces new events whose progression cannot be further controlled. Naturally, this principle can also be found in most of the other procedures of algorithmic composition that generate a particular musical structure on the basis of a rule system of any kind. The formulation of such rules, as e.g. in the framework of a rewriting system, however, enables a high predictability of the musical results and therefore also a realization of specific musical concepts. But it is exactly this complex behavior of deterministic chaotic systems which may present a high incentive to apply them to tasks of algorithmic composition: "The great attraction of non-linear dynamical systems for compositional use is their natural affinity to the behaviors of phenomena in the real world, coupled with the mechanical efficiency of their computation and control. Chaotic systems offer a means of generating a variety of raw materials within a nonetheless globally consistent context. Chaotic sequences embody a process of transformation, the internal coherence of which is ensured by the rules encoded in the equations." [1].

Lindenmayer systems are rewriting systems like generative grammars and generate symbol strings by applying production rules; originally, they were developed for the simulation of growth processes. In contrast to generative grammars where, in general, the desired output is only produced after all substitutions have been carried out, in a traditional L-system single terminals are replaced by a larger number of terminals. Therefore, after every substitution process symbols are available for musical mapping. As a consequence, on the one hand, the number of produced symbols typically increases considerably in every generation[26]; on the other hand, the L-system is also subject to a temporal development and may therefore also reflect process-like compositional approaches. Due to this basic trait, a Lindenmayer system is also very well suited to representing self-similar structures and enables the effective mapping of fractal concepts on musical parameters.

The representation of the musical interpretation of the results of an LS is of major significance. Strategies that are based on a musical interpretation of turtle graphics, i.e. using an already performed mapping as the starting point for a further mapping, are, however, in general not able to fully reflect the specific behavior of an LS in the musical mapping. The work of DuBois shows a number of alternative strategies in which different formalisms of LS are used for musical mapping. It is exactly this diversity of possible Lindenmayer system forms that opens up a wide field of musical structure genesis to algorithmic composition. The possible applications of the generation of musical structure may range from the simple musical representation of self-similarity to the production of complex musical structures in the framework of contemporary musical creation.[27]

For most of the abovementioned approaches the notion of scale is an important aspect of musical structure generation. This term gets an interesting new dimension by Mandelbrot, who distinguishes between scaling and scalebound structures [13]. Scalebound structures, whether they are artificial or natural, are characterized by a few dominating elements of scale, whereas scaling structures consist of a large amount of different scales, without any dominating elements being capable of characterizing the whole structure. This distinction could also be helpful for the classification of musical structure, which can be created by some dominating formal principles, such as a strict self-similar principle (in analogy to a scalebound object), or arises from various decisions, stochastic principles and the like (in analogy to a scaling object). Often, the character of a composition changes during the various stages of creation: Starting as a "scalebound object" by a formal principle, the resulting structure will be altered according to personal aesthetical preferences and by this transforms little by little into a "scaling object." Besides the various structuring elements of a composition, a "successful" piece of music will always show the phenomenon of emergence[28] and therefore seems to overcome the restrictions of its formal principles, or, as Mandelbrot states with an example from the visual arts: "Incidentally, while the maximum size of a painting is that of the whole canvas, many painters succeed in giving the impression that it contains depicted objects of larger sizes. Similarly, a brush stroke determines the minimum size, but many portrait jewels seem to include detail that is known to be physically impossible to depict at the scale depicted." [13, p. 46, 47].

[26] This property may easily be modified by appropriate rewriting rules, cf. the LS presented by DuBois, which produces symbol strings of the same length in every generation and equals a 1-dimensional cellular automaton.

[27] So, e.g. the contemporary composer Hanspeter Kyburz developed complex formalisms of LS for musical structure genesis, cf. [24].

[28] Cf. the section "Agents" in chapter 10.

References

1. Bidlack R (1992) Chaotic systems as simple (but complex) compositional algorithms. Computer Music Journal, 16/3, 1992
2. Bolognesi T (1983) Automatic composition: experiments with self-similar music. Computer Music Journal, 7/1, 1983
3. Dodge C (1988) Profile: a musical fractal. Computer Music Journal, 12/3, 1988
4. Dodge C, Jerse TA (1997) Computer music: synthesis, composition, and performance, 2nd edn. Schirmer Books, New York. ISBN 0-02-864682-7
5. DuBois RL (2003) Applications of generative string-substitution systems in computer music. Dissertation. Columbia University, 2003
6. Gleick J (1987) Chaos: making a new science. Penguin Books, New York. ISBN 0-14-00 9250-1
7. Gogins M (1991) Iterated functions systems music. Computer Music Journal, 15/1, 1991
8. Hogeweg P, Hesper B (1974) A model study on biomorphological description. Pattern Recognition, 6, 1974
9. Lang B (1996) Diminuendo. Über selbstähnliche Verkleinerungen. In: Beiträge zur Elektronischen Musik, 7. Institut für Elektronische Musik (IEM) an der Universität für Musik und darstellende Kunst in Graz, Graz.
10. Leach J, Fitch J (1995) Nature, music, and algorithmic composition. Computer Music Journal, 19/2, 1995
11. Lindenmayer A (1968) Mathematical models for cellular interaction in development. Journal of Theoretical Biology, 18, 1968
12. Mandelbrot B, Van Ness J (1968) Fractional brownian motions, fractional noises and applications. SIAM Review, 10/4
13. Mandelbrot B (1981) Scalebound or scaling shapes: A useful distinction in the visual arts and in the natural sciences. Leonardo, 14, 1981
14. Mandelbrot B (1982) The fractal geometry of nature. W. H. Freeman and Company, New York. ISBN 0-7167-1168-9
15. Mech R (2004) CPFG Version 4.0 User's Manual based on the CPFG Version 2.7 User's Manual by Mark James, Mark Hammel, Jim Hanan, Radomir Mech, Przemyslaw Prusinkiewicz with contributions by Radoslaw Karwowski. http://algorithmicbotany.org/lstudio/CPFGman.pdf Cited 11 Nov 2004
16. McCormack J (1996) Grammar based music composition. In: Stocker R, Jelinek H, Durnota B, Bossomaier T (eds) (1996) Complex systems 96: From local interactions to global phenomena. ISO Press, Amsterdam. ISBN 9-05-199284-X
17. Peitgen HO, Richter PH (2001) The beauty of fractals. Images of complex dynamical systems. Springer, Berlin. ISBN 978-3540158516
18. Poincaré H (1952) Science and method. Dover Publications, New York. ISBN 10 0486602214
19. Pressing J (1988) Nonlinear maps as generators of musical design. Computer Music Journal, 12/2, 1988
20. Prusinkiewicz P (1986) Score generation with L-systems. In: Proceedings of the 1986 International Computer Music Conference. International Computer Music Association, San Francisco
21. Prusinkiewicz P, Lindenmayer A (1990) The algorithmic beauty of plants (The Virtual Laboratory). Springer, New York. ASIN 0387972978
22. Prusinkiewicz P (2004) Algorithmic Botany. http://algorithmicbotany.org/ Cited 8 Nov 2004
23. Smith AR (1984) Plants, fractals, and formal languages. Computer Graphics, 18, 3 July, 1984
24. Supper M (2001) A few remarks on algorithmic composition. Computer Music Journal 25/1, pages 48–53, 2001
25. Voss RF, Clarke J (1978) "1/f noise" in music: Music from 1/f noise. Journal of the Acoustical Society of America, 63/1, 1978
26. Yorke JA, Li TY (1975) Period three implies chaos. The American Mathematical Monthly, 82/10, 1975

Chapter 7
Genetic Algorithms

Genetic algorithms as a particular class of *evolutionary algorithms*, i.e. strategies modeled on natural systems, are stochastic search techniques. The basic models were inspired by Darwin's theory of evolution. Problem solving strategies result from the application of quasi-biological procedures in evolutionary processes. The terminology of genetic algorithms including "selection," "mutation," "survival of the fittest," etc. illustrates the principles of these algorithms as well as their conceptual proximity to biological selection processes.

In the initial stages of their development, these principles took shape in two different models: From the 1960s on, Ingo Rechenberg and Hans-Paul Schwefel introduced the *evolution strategies*[1] at the Technical University of Berlin, and in the 1970s, the Americans John H. Holland[2] and David E. Goldberg[3] developed genetic algorithms. Rechenberg and Schwefel's models are based upon a graphic notation[4] and were modeled on biological procedures for the development of technical optimization techniques. Holland and Goldberg's genetic algorithms use the principles of coding and transmission of data in biological systems for modeling search strategies. These two approaches developed, to a great extent, separately from each other. For application in music, the problem solving strategies of the "American school" are applied, and for this reason, Rechenberg's model will not be explained here in detail.

7.1 The Biological Model

DNA in a cell consists of *chromosomes* that are made up of *genes*. Genes describe amino acid sequences of *proteins* and are responsible for the development of dif-

[1] "Optimierung technischer Systeme nach Prinzipien der biologischen Evolution" [36].

[2] "Adaption in Natural and Artificial Systems" [17].

[3] "Genetic Algorithms in Search, Optimization, and Machine Learning" [16].

[4] An illustration of the graphic symbols and an introductory overview can be found in [38].

Fig. 7.1 John Holland and David Goldberg. With kind permission of John Holland and David Goldberg.

ferent traits that become manifest in different ways by transferring genetic information. The total complement of genes is referred to as a *genome*. The entirety of an individual's hereditary information is known by the term *genotype* and the specific manifestation of his or her features called a *phenotype*. Genetic variability is ensured by a population with differing genetic characteristics as well as a continuous adaptation to changing environmental conditions. Genetic variations are caused by a process called *meiosis*, by which the hereditary disposition of the parents is allocated differently to the cells off the offspring, as well as by *mutation* of the genes, chromosomes or the whole genome. According to Darwin's theory of evolution,[5] the competing behavior of living organisms promotes the passing on of the genetic information of the fittest, meaning those organisms best able to survive in a particular environment. Consequently, this leads to the *survival of the fittest*, a term which can also be found in the terminology of genetic algorithms as the fitness function.

7.2 Genetic Algorithms as Stochastic Search Techniques

Genetic algorithms, which model the evolutionary processes in computer simulation, are methods that are used to solve search and optimization problems. For the application of a genetic algorithm, domain-specific knowledge of the problem to be solved is not necessary. Therefore, this class of algorithms is especially suitable for tasks that are difficult to model mathematically or for problem domains that do not have an explicit superior rule system.

By analogy to the biological model, the respective computer program serves as the habitat that provides particular conditions for surviving and heredity. In this artificial living space, populations of individuals, or chromosomes, are produced whose adaptation to an objective, referred to as *objective score*, is examined by means of

[5] "On the Origin of Species by Means of Natural Selection, or the Preservation of Favoured Races in the Struggle for Life" [8].

a *fitness function*. The fitness function may represent a mathematical function, a comparison set, or a rule-based system that examines the ability of a chromosome to fulfill the objective score. In algorithmic compositions, human fitness-raters are also frequently used; this approach, however, is subject to some restrictions (see below). In a simulation, binary coded symbol strings are generally applied to represent an individual. New populations are produced with the principles of *crossover* and *mutation* of chromosomes. In crossover, which may happen in several different ways, corresponding bits of the parent strings are swapped to produce new chromosomes. In order to guarantee the genetic diversity of a new population, in addition to crossover, mutation is applied which modifies elements of the symbol strings in various ways by using stochastic procedures. Consequently, in a binary representation, an arbitrary piece of the string could be changed from 0 to 1 or from 1 to 0. The fittest individuals are selected in each new run for crossover and mutation or passed on without being modified. This whole process continues until a chromosome is generated that best matches the objective score. In this context, artificial molecules may serve as an example in which new combinations are generated by swapping components; in every new generation, these new molecules are examined for their adequacy to an objective score, such as, for example, strength.

In principle, the scheme of a genetic algorithm is structured as follows:

1. Generate random starting population of n chromosomes
2. Calculate the fitness of each chromosome If good enough, dump the result, END Else:
3. The fittest chromosomes are transferred unmodified or undergo crossover or mutation
4. Select a number of fittest chromosomes as starting individuals
5. Create next generation and repeat from step 2

In "Genetic Algorithms in Search, Optimization and Machine Learning" [16], David Goldberg describes an example [16, p. 7–19] of a simple application of a genetic algorithm. Consider a black box system which produces a numeric output for different positions of its switches (figure 7.2). When the switches are represented as binary symbol sequences (chromosomes), numerical values result in particular positions that may also be used to determine a simple fitness function: The higher a chromosome's numeric value, the better its fitness.

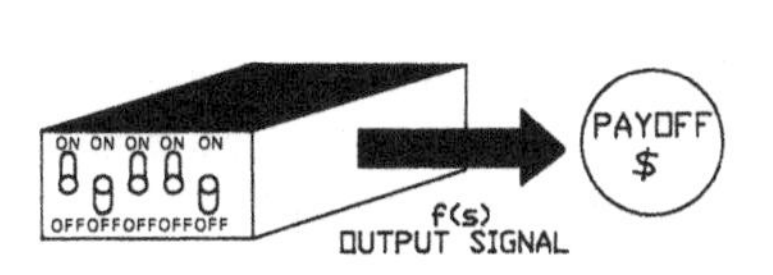

No.	String	Fitness	% of Total
1	01101	169	14.4
2	11000	576	49.2
3	01000	64	5.5
4	10011	361	30.9
Total		1170	100.0

Fig. 7.2 Black box, output and fitness values (rightmost is % of sum of total fitness) [16, p. 8, 11].

Now, every chromosome's fitness is allotted (according to Goldberg) to the segments of a so-called "roulette wheel" (figure 7.3) in order to select chromosomes that may be used for crossover by creating weighted probabilities:

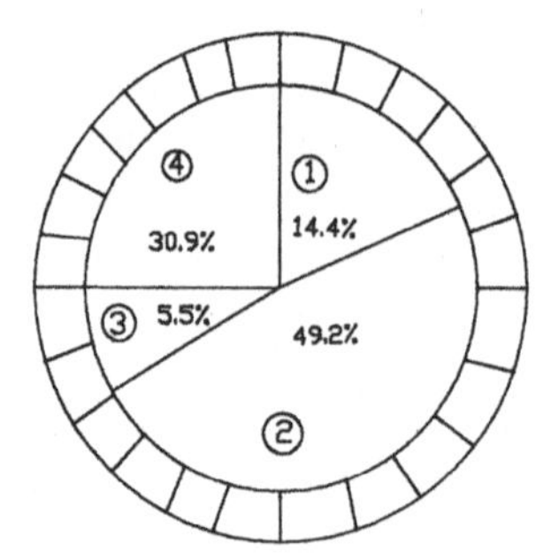

Fig. 7.3 Probabilities for the selection of chromosomes in crossover [16, p. 11]. © 1989 Pearson Education Inc. Reproduced by permission.

Because the search space is restricted by binary symbol strings to a value between 0 0 0 0 0 and 1 1 1 1 1, each chromosome x has a possible value between 0 and 31. The function underlying the black box is assumed to be $f(x) = x^2$. Suppose a population consists of four chromosomes (strings) whose initial configuration may be acquired by flipping a coin twenty times. In the following example, the values of the black box are applied. After generating the initial population, values as given in figure 7.4 result; the columns 3 to 6 indicate the following:

- x : Value of the binary string in decimal notation.
- $f(x) = x^2$: Value of the chromosome in regards to the fitness function.
- $\frac{f_i}{\sum f}$: Relation of the single values to the total sum.
- $\frac{f_i}{\bar{f}}$: Relation of the single chromosome values to the average of all chromosomes.
- Actual count from Roulette Wheel: The results of the roulette wheel with regard to the selection of the chromosomes used for crossover.

The production of a new population starts with the selection of chromosomes, to which crossover will be applied. In this example (figure 7.4), chromosome 1 and chromosome 4 are selected once, chromosome 2 is selected twice, and chromosome 3 is not selected at all. These chromosomes (first column in figure 7.5) are now crossed-over with randomly selected chromosomes (second column in figure 7.5) at a randomly chosen crossover point and result in new chromosomes; this process is referred to as *one-point crossover*. As a result of this crossover, the sum, the average, and the highest achieved value of single chromosomes have increased in regard to the fitness function. Dispensing with mutation, this example shows some basic operations of genetic algorithms that, however, may be expanded upon in different ways.

String No.	Initial Population (Randomly Generated)	x Value (Unsigned Integer)	$f(x)$ x^2	pselect$_i$ $\frac{f_i}{\Sigma f}$	Expected count $\frac{f_i}{\bar{f}}$	Actual Count from Roulette Wheel
1	0 1 1 0 1	13	169	0.14	0.58	1
2	1 1 0 0 0	24	576	0.49	1.97	2
3	0 1 0 0 0	8	64	0.06	0.22	0
4	1 0 0 1 1	19	361	0.31	1.23	1
Sum			1170	1.00	4.00	4.0
Average			293	0.25	1.00	1.0
Max			576	0.49	1.97	2.0

Fig. 7.4 Chromosome values regarding fitness and selection for crossover [16, p. 16]. © 1989 Pearson Education Inc. Reproduced by permission.

Mating Pool after Reproduction (Cross Site Shown)	Mate (Randomly Selected)	Crossover Site (Randomly Selected)	New Population	x Value	$f(x)$ x^2
0 1 1 0 | 1	2	4	0 1 1 0 0	12	144
1 1 0 0 | 0	1	4	1 1 0 0 1	25	625
1 1 | 0 0 0	4	2	1 1 0 1 1	27	729
1 0 | 0 1 1	3	2	1 0 0 0 0	16	256
					1754
					439
					729

Fig. 7.5 New population after one crossover [16, p. 17]. © 1989 Pearson Education Inc. Reproduced by permission.

7.3 Genetic Programming

An approach that applies functions instead of symbol strings and therefore does not generate values, but rather programs as a result of the genetic operations, is described by John Koza in "Genetic Programming: On the Programming of Computers by Means of Natural Selection" [24] published in 1992. In this application of genetic principles to the development of computer programs, the gene pool's point of origin generates a number of domain-specific functions and variables. In the generation of populations, function calls or variables (or terminals representing a single fixed value) are either randomly arranged in a tree structure or functions and variables are randomly assigned to a tree of a particular depth (figure 7.6). Once the position of the variables and functions enables the processing of the program, an output is produced that is examined by means of a fitness function. The root of the graph is usually assigned with a function; crossover occurs with two chromosomes by randomly switching nodes and their children, as shown in figure 7.7.

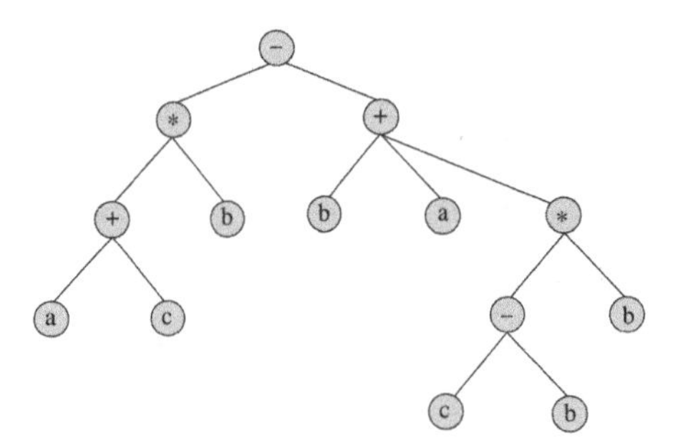

Fig. 7.6 Tree structure made up of functions and variables.

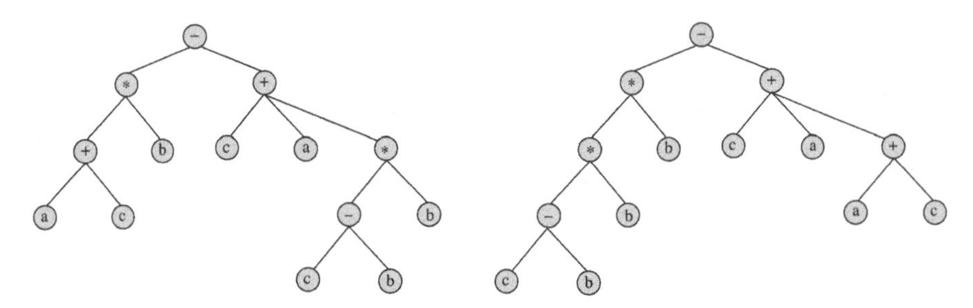

Fig. 7.7 Crossover in genetic programming.

Mutation replaces functions or variables of a random position by randomly generating a tree structure, a single function or a variable. As an example of a more complex application of genetic programming, Koza described genetic programming as used in the simulation of a multiplexer. A multiplexer is a combinatorial circuit with multiple inputs and one output. On the input side there are n control signals, also called address bits, and 2^n inputs. The address bits determine which of the input signals goes to the output. Koza used a multiplexer with three address bits (A0–A2) and eight input signals accordingly (D0–D7), as shown in figure 7.8. Genetic programming is, in this case, intended to generate a program that is able to correctly simulate the operation mode of the multiplexer with logical functions and terminals. However, it is assumed that the operation mode of the multiplexer is unknown; so a program structure is generated without domain-specific knowledge. This task is especially suitable for genetic programming, because in this example, a blind search is hardly feasible as a problem solving procedure due to the enormous search space of all possible combinations of the functions and terminals.

Koza used the functions AND, OR, NOT, IF dividing IF into IF, THEN, ELSE, and the terminal alphabet consisting of A0, A1, A2, D0, D1, D2, D3, D4, D5, D6,

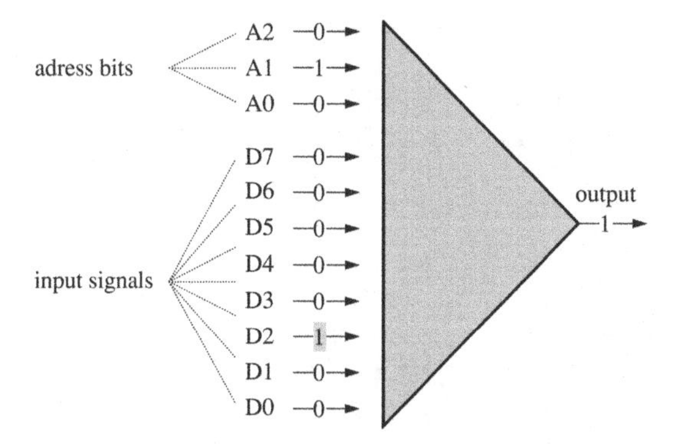

Fig. 7.8 Scheme of a multiplexer.

D7. The program is built up with a nested list structure of functions and terminals in Lisp syntax.[6] The fitness function results from the total of all correct circuits of the multiplexer that may be indicated on the input side by combining the terminals with 2^{11} possible circuits. The number of the valid solutions (2048) is taken as fitness for the generated programs. For the size of the population a value of 4000 is assumed and at most, 51 generations are produced. In generation 0, for example, the following programs with attributed fitness are generated:

(IF (IF (IF D2 D2 D2) D2 D2) D2 D2): no output making sense
(OR (NOT A1) (NOT (IF (AND A2 A0) D7 D3))): 768 from 2048 possible solutions

In generations 4 and 5, 1664 correct solutions can already be found. One of these functions of generation 5 is:
(IF A0 (IF A2 D7 D3) (IF A2 D4 (IF A1 D2 (IF A2 D7 D0))))

In order to represent the problem solving capacity of the programs, a graphic illustration is used that shows all possible circuits on the input side of the multiplexer as squares. The state of these squares indicates whether the correct bit goes to the output (full = correct, empty = false output signal). Figure 7.9 compares examples of best-of functions of the generations 0 and 4.

Finally, in generation 9 the function:

(IF A0 (IF A2 (IF A1 D7 (IF A0 D5 D0))
(IF A0 (IF A1 (IF A2 D7 D3) D1) D0))
(IF A2 (IF A1 D6 D4)
(IF A1 D2 (IF A2 D7 D0)))))

[6] At the beginning of the brackets is the function, followed by arguments, e.g. (/(+ 1 2 3 (* 2 3))2) equivalent: $\frac{1+2+3+(2*3)}{2}$. This notation is also called *prefix notation*.

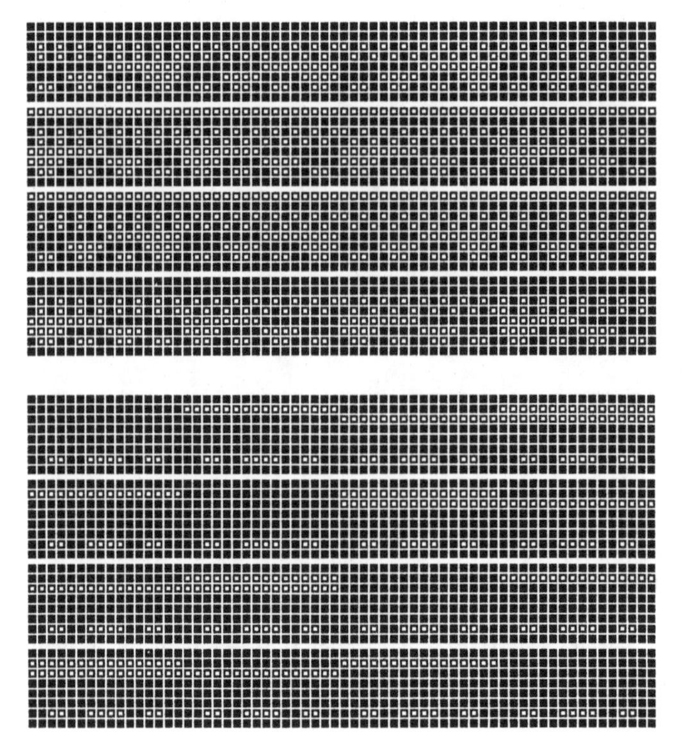

Fig. 7.9 Best-of function of the generations 0 (top) and 4 (bottom) [24, p. 175, 178].

detects 2048 correct solutions. A simplified form of the program may be transferred to the following form either manually or with an editing function:

```
( IF A0 ( IF A2 ( IF A1 D7 D5) ( IF A1 D3 D1 ) )
( IF A2 ( IF A1 D6 D4 ) ( IF A1 D2 D0 ) ) )
```

7.4 Genetic Algorithms in Algorithmic Composition

As one of the first applications of genetic processes in algorithmic composition, Andrew Horner and David Goldberg [18] described the generation of melodic material by means of "thematic bridging." This technique modifies a start pattern using a number of functions and compares the results with another pattern as fitness function. For the generation of a composition, Horner and Goldberg used six such cycles in order to produce melodic material that is then further structured by a five-voice canonical layering. The functions that in this case undertake the tasks of crossover and mutation are, for example, with an initial pattern of Gb, Bb, F, Ab, Db and a reference pattern of F, Ab, Eb, as follows:

1. Start pattern: Gb Bb F Ab Db
2. Deletion of the last element: Gb Bb F Ab
3. Random swapping of the elements: Bb F Ab Gb
4. Deletion of the last element: Bb F Ab
5. Modification of the first element: Eb F Ab
6. Random swapping of the elements: F Ab Eb

Output: Gb Bb F Ab Bb F Ab Gb Bb F Ab Eb F Ab F Ab Eb.

In a two-stage process, the fitness function compares the conformity of the tone pitches of the output with the pitches in the reference pattern, as well as the length of the output with a desired objective. Although Horner and Goldberg's procedural method is very simple in this example, it describes the possibility of arbitrarily adjusting the principles of crossover, mutation and fitness evaluation with regard to the structuring of musical material.

7.4.1 Analogies to the Process of Composition

In their applications of genetic algorithms to the generation of musical structure, Bruce L. Jacob [20, 21] and Andrew Gartland-Jones [11, 12], with Peter Copley [13] emphasized this procedure's similarity to a traditional compositional process. Jacob outlines the objective of Variations, his algorithmic composition system: "The system was designed to reproduce very closely the creative process that this author uses when composing music [...]," [21, p. 2] and Gartland-Jones refers to the general function principle of a genetic algorithm as follows: "A commonly used compositional process may be described as taking an existing musical idea and changing it in some way. Musicians in various styles and genres may follow the process differently, some through improvisation, others through pencil and paper, but what is most often practiced is taking an existing idea and *mutating* it to provide a new idea. In fact, mutation is closely related to notions of development, which lie at the heart of western musical concepts of form and structure. It may even be possible to see development as *directed* mutation. [...] With the core elements of GA's being mutation (including cross-over) and fitness assessment, there appears to be a strong correlation with at least some aspects of the human process of generating musical material." [11, p. 2–3].

According to Gartland-Jones, populations of phrases of two bars obtain their fitness after the application of structure-modifying operations by a simple comparison with phrases of a given corpus. This principle finds further application in the generation of obbligatos, interactive installations and a software system. Distinctive parameters for notes are pitch, duration and velocity; the populations are generated considering both the key and mode of the reference patterns. Ten functions are selected to serve as genetic operations that, among other things, mirror, invert, rearrange and mutate the material. The comparison of the patterns is applied on every note: If a note corresponds to a note of the reference pattern, it obtains fitness "1." Consequently, the fitness of a pattern is the total of the fitness of all single notes, whereas the position of the notes in the pattern is not taken into account. For the application of this principle in an installation [11, p. 9–10] the space is divided into sixteen sectors that are assigned four cycles with different fitness. Sensors capture the spatial position of the visitors and display the corresponding patterns with speakers. Obbligatos are generated in another application of this principle in relation to a user's input. Here, each of the last four bars of incoming MIDI data are subject to different genetic operations and made audible simultaneously. Implementations of this principle within a software program[7] make use of virtual building blocks that interact musically. Each of these blocks has its own musical pattern, and is able to send this pattern to other blocks and also receive patterns from these. When a pattern is passed to another block, it serves as a reference pattern for the pattern of the receiving block. The latter generates a new musical pattern based on its own music and the music it has just been passed, which again may be sent.

In his software project Variations, Jacob used three program modules for the generation of musical structure by means of genetic algorithms. The basic components of his software are the modules "Composer," "Ear" and "Arranger." In the "Composer," variations of existing motifs are produced with the help of different techniques. The "Ear" evaluates the correctness of the motifs by permitted interval combinations and produces musical phrases that are assembled by the "Arranger" into larger units, which will be finally evaluated by a user. At the beginning, the user initializes the "Composer" with a number of motifs to which the module adds further motifs through transposition or rhythmic changes, for example. The evaluation of the motifs in the "Ear" is performed automatically by applying a simple comparison of patterns in order to examine whether all newly generated motifs show the same intervallic relations as the entered motifs. Because a motif in this system may be composed of both single notes and chords, in the evaluation the horizontal as well as the vertical intervallic relations are equally important. A valid new motif is at least a subset of the intervallic relations of the entered motifs; similarly, the doubling of chord components is allowed, as is the transposition of valid new generations or their rhythmic changes. Figure 7.10 illustrates this principle: In the upper part, there is an entered motif; in the lower left part, there are some new generations evaluated as valid; in the lower right part, some evaluated as invalid.

The "Ear" arranges all valid motifs into phrases that are composed by the "Arranger" by means of combination to larger musical units which are finally rated by the user.

---

[7] MusicBox [12]; in a further development with Peter Copley: IndagoSonus [13].

Fig. 7.10 Motif (top) and correct (bottom left) and incorrect (bottom right) new generation.

7.4.2 Varieties of Genetic Operations and Fitness Ratings

George Papadopoulos and Geraint Wiggins [29] described the generation of melodies of variable length and rhythmic structure over given chord sequences. Domain-specific genetic operations work with symbolically encoded symbol strings that must undergo an algorithmic fitness evaluation. A fitness evaluation done by a subject, also referred to as *human fitness function*, is not considered in this approach as a listener's preferences may change over time and the large number of evaluations brings about symptoms of fatigue as well. The need to evaluate all generations of a genetic algorithm – also known as *fitness-bottleneck* – is a basic problem occurring in evaluations performed exclusively by a user. Alternative strategies may either keep the number of populations low or, as in Jacob's example, make use of multi-stage fitness evaluations applying either algorithmic or human evaluations for the respective tasks.

Papadopoulos et al. encoded tone pitches in relation to an underlying chord, whereas a second entry determines the duration. Instead of the pitch, a symbol for a pause may also be placed. As a frame, 21 pitches are taken into account based on seven and eight note scales. The initial populations are chosen randomly, whereas a pause is placed instead of a pitch with a 12.5 % probability. In the application of genetic operations, classes of functions are used: "Local mutations" operate on a fragment of a chromosome of variable length by means of transpositions, inversions, ascending and descending arrangement of the elements etc. "Copy and operate mutations" copy or swap randomly chosen fragments of the chromosome. "Restricted copy and operate mutations" also swap and copy segments of the symbol strings but operate on downbeats in order to establish motivic structures at easily recognizable fragments. Within the automatic fitness evaluation, eight categories of criteria are applied that examine the chromosomes for interval sizes within the melodic progressions, correct functional harmonic context and permitted notes on downbeats, among other things. In addition, another class of functions produces a comparison of patterns based exclusively on tone pitches with a musical reference corpus. Figure 7.11 shows a production of Papadopoulos and Wiggins's system.

One application of genetic algorithms as used for the harmonization of given melodies is realized by Ryan Mc Intyre with his "Bach in a Box" system [26]. As a restriction, only progressions in C major are produced; secondary dominants, pivot chords and key changes are ignored. The progressions are represented in a number sequence in which four values are assigned to a chord (1st value: soprano, 2nd value: alto, 3rd value: tenor, 4th value: bass). A starting population is generated by

Fig. 7.11 A melody generated by the system of Papadopoulos and Wiggins.

randomly choosing number values from 0 to 28 for every second to fourth position (alto to bass) so that the result is the voices from alto to bass lying in a diatonic range of four octaves. Crossover is applied in the usual way; in the mutation, randomly selected segments of the symbol string are swapped. Fitness is evaluated for every single note in regard to musical parameters whose fulfillment is rated with different weightings. These rules describe some rudimentary principles for a style typical four-voice chorale movement, such as permitted vocal ranges and possible chord tone doublings. Fitness is rated in a three-stage model in which a certain percentage of fitness of a particular category must always be reached in order to be considered for the next level. So, for example, a chromosome must receive at least 85% of its possible fitness from the criteria "good chords," "good ranges," "good Start-Stop" (suitable input and output chord) as well as "harmonic interest" (the harmonic variability) before it can receive fitness from a final fitness evaluation. It becomes evident that the three-stage model is better, but is nevertheless not able to achieve optimal fitness even in a very large number of generations.

An interesting aspect in the work of Michael Towsey, Andrew Brown, Susan Wright, and Joachim Diederich [43] is the application of genetic algorithms to already existing melodic material by means of a multi-dimensional fitness function. Furthermore, these authors made use of a number of statistical studies on the generation of a comprehensive fitness function. The starting material consists of 28 melodies of composers from the Renaissance to the Classical period as well as a number of children's songs and popular music themes. Towsey et al. attempted to establish an objective standard for a good melody by means of a number of statistical features. Each melody is examined for 21 criteria including used dissonances, pitch variety and numerous others. The results make it possible for the authors to generate a multi-dimensional fitness function which may also be used to classify the starting melodies. For each melody, every fitness dimension results in a value between 0 and 1, the number of occurrences of a particular feature being related to its possible number of occurrences. According to this, a melody with ten notes, of which five tone pitches are different, for example, receives a fitness measure of 0.5 with regard to the criterion "pitch variety." For all melodies, criteria that show the smallest deviations regarding an average value – meaning the melodies are similar to each other in this one aspect – seem to represent general quality criteria and are respectively rated higher in terms of fitness. The following fitness dimensions show the lowest deviations:

1. Pitch variety: $\frac{\Sigma \textit{distinct tone pitches}}{\Sigma \textit{tone pitches}}$
2. Dissonant intervals: $\frac{\Sigma \textit{interval dissonances}}{\Sigma \textit{intervals}}$

   The authors indicate the following "dissonance ratings:"

Interval	Dissonance rating
0 1 2 3 4 5 7 8 9 12	0.0
10	0.5
6 11 13	1.0

3. Contour direction: $\frac{\Sigma \textit{rising intervals}}{\Sigma \textit{intervals}}$
4. Contour stability: $\frac{\Sigma \textit{rising intervals moving in same direction}}{\Sigma \textit{intervals - 1}}$
5. Rhythmic variety: $\frac{\Sigma \textit{distinct rhythmic values}}{16}$

   (The denominator is the same for each melody because the corpus only consists of sixteen different rhythmic values.)

6. Rhythmic range: $\frac{\Sigma \textit{(max note duration) - (min note duration)}}{16}$

   The means and standard deviations of these features are as shown below:

Category	Mean	Standard deviation +, –
1. Pitch Variety	0.27	0.11
2. Dissonant Intervals	0.01	0.02
3. Contour Direction	0.49	0.06
4. Contour Stability	0.40	0.11
5. Rhythmic Variety	0.24	0.07
6. Rhythmic Range	0.32	0.11

This multi-stage fitness evaluation also allows for a classification of the corpus within the three classes "Classical," "Early" and "Nursery." The results of the analysis also prove true through a *principal component analysis*[8] applied comparatively as well as a following *clustering*[9] that shows, for example, all melodies of the musical style "Classical" within three clusters. Of course, it can be argued that the approach of Towsey et al. works with a corpus that is too small; the authors themselves also point out that this problem may lead to difficulties in regard to a consistent comprehension of a given style. If the analysis methods enable distinctions of certain styles, this aspect will also naturally show the inhomogeneity of the corpus.

Whether it is possible to find quality criteria for "good" melodies that are binding beyond style remains to be seen; a particularly successful melody – considering the diverse ways of interpreting "good," "successful," etc. – may also show naturally in a large derivation from a particular musical criterion. Nevertheless, the authors work is proof of the particular suitability of multi-stage fitness functions for musical resynthesis because, here, characteristics of the corpus may be looked at from several different perspectives.

7.4.3 Limits of Genetic Algorithms – A Comparison

Somnuk Phon-Amnuaisuk, Andrew Tuson and Geraint Wiggins[10] used genetic algorithms to harmonize soprano voices. In this approach, genetic operators such as mutation of chord type, swapping voices etc. are applied and the fitness functions are executed in compliance with basic rules of music theory. Figure 7.12 shows a harmonization performed using this method. The authors work is interesting also due to the fact that they point out intrinsic weaknesses of a genetic algorithm regarding the performance of particular tasks. An optimal fitness for all categories cannot be achieved even within a very large number of productions; the generation of chord progressions turns out to be extremely difficult.

Fig. 7.12 Harmonization generated by the system of Phon-Amnuaisuk et al.

First, harmonic progressions are highly context-sensitive so that when changing a chord, the functional context of the entire environment must be recreated. In the

[8] In this analysis method, a large number of variables are reduced to a smaller number of influencing values. Variables that intercorrelate to a large extent are combined.

[9] Formation of disordered material in categories or groups by means of algorithms; see chapter 10.

[10] In [33] and with George Papadopoulos in [45] – here, the production of a melody voice is treated in addition, like in [29].

context of a genetic algorithm this may mean that when improving a particular fitness dimension, the newly generated chromosome provides worse values for other fitness dimensions. Another problem is rooted in the fact that the genetic algorithm yields good results in small musical tasks without, however, being able to generate the musical part as a whole in a way that makes sense. Finally, consideration must be given to the fact that the search in the state space is heuristic and not complete, and this, as a result, is a reason for the impossibility of always reaching an optimal solution for the musical task.

If there exists sufficient problem-specific knowledge in a given musical domain, then rule-based systems are in most cases superior to a genetic algorithm for the abovementioned reasons, as described by Phon-Amnuaisuk and Wiggins in a further study [34]. In this comparison, harmonization is generated by a genetic algorithm as well as by a rule-based system; the same domain knowledge is made available to both techniques. Knowledge of correct progressions is implemented in the genetic algorithm by the type of data representation, the genetic operations and the fitness evaluations. The structuring of the genetic algorithm in the harmonization of a given soprano voice is done similar to the above approach. The rule-based system first produces chord progressions that serve as a basis for the further arrangement of the musical fine structure. A backtracking system[11] controls the single steps of the generation process. Evaluation is done by means of giving penalty points according to the breaking of simple rules of music theory – for the abovementioned reasons, as expected, the rule-based system yields better results. Figure 7.13 displays the results of the harmonization processes of both methods, the best result from the genetic algorithm scoring twenty penalty points after 250 generations, the rule-based system, however, scoring only two penalty points in one variant.

Fig. 7.13 Comparison of different automatic harmonizations by means of a genetic algorithm (left) and a rule-based system (right).

The problems regarding voice leading occurring in the example of the genetic algorithm are caused by the parallel octave at the end of the first bar as well as by the last pitch in the bass part being out of range. In the result from the rule-based system, the generation of the progressions "V–ii" as well as "V–V^7" is considered inappropriate.

[11] In case a solution is not satisfactory, the solution found before is calculated anew; this process is applied until a satisfactory result is reached.

Even if the rule-based system shows a clear advantage over the genetic algorithm in this study, the fact that the rule-based approach is also subject to strong restrictions must be considered. The system is determined by the implemented rules; the output is completely foreseeable so that in some cases the gain of any insight from musical resynthesis must also be questioned. On the other hand, the genetic algorithm is completely able to produce surprising and yet musically satisfying results. Most of the algorithmic fitness functions generalize musical information by means of knowledge-based or rule-based strategies in order to be able to represent a uniform set of criteria for the system's outputs. In the recognition of inventive solutions – that are actually violations of musical convention – these fitness functions reach their limits early. In addition to the abovementioned strategies, artificial neural networks may also be applied for fitness evaluations; some works on this subject were produced by Lee Spector and Adam Alpern, Brad Johanson and Al Biles (see below). A neural network may produce surprising results, both as a producing and an evaluating entity. Regarding context-sensitive structure, however, neural networks are often subject to the same restrictions as genetic algorithms so that in these cases, rating by a user is recommended.[12]

In order to reduce the enormously increasing search space in these cases, Paul Pigg [35] suggested a two-staged model for the improvement of these restrictions. In his approach, the user structures movements such as Intro, Chorus, Solo, etc. and initializes them by indicating bar and key. Two genetic algorithms then generate the fine structure: The first genetic algorithm produces a genetic pool of bars that possess all characteristics of the corresponding movement; by this, a separate population is generated for each movement. The chromosome is represented by two separate symbol strings, the first referring to the pitch class and the other to the position of the octave. In place of a pitch class, symbols for rests and holds may be introduced in the symbol string as well. Crossover and mutation find application as genetic operators, whereas crossover is carried out in the usual way in the pitch classes and octaves. Mutation processes consist of shortening, extending and changing notes as well as random octave mutation. Fitness is rated by means of the simple principle that pitch classes are compared with the scale degrees that are, in turn, determined by the key of the respective movement. The evaluation of every single octave is based on the mean value of the chromosome; each derivation reduces an optimal fitness. The second genetic algorithm generates further variations on the basis of these chromosomes. In contrast to the first algorithm, this one generates chromosomes of greater length and its fitness function additionally includes triads belonging to the scale in the evaluation. Even when, due to the simple comparison of patterns, the fitness functions in Pigg's model have difficulties in recognizing a "coherent" melodic structure, the pre-structuring of the musical material is an interesting approach to an efficient reduction of a large search space.

[12] The abovementioned works are treated in the sections on genetic programming and interactive systems; for further approaches using neural networks as fitness functions, see [37, p. 2–3].

7.4.4 Rhythmic Generators

So far, applications regarding the generation of melodic and harmonic structure have been described; however, genetic algorithms also find application in the production of rhythm. Damon Horowitz developed a system [19] in which rhythmic patterns of differently instrumented structure are generated by means of the preferences and ratings of a user. Each chromosome represents a bar whose rhythmic values may comprise pauses as well as note values between 1/16 triplets and 1/4 notes. A number of rhythmic parameters including density, repetition, stressed beats and many more are created by the user in different ways according to their importance for the generation of the populations. Each of these criteria is furthermore assigned an optimal value for the fitness evaluation that must be reached, as well as a weighting in relation to the other criteria. This weighting controls how dominantly the respective criterion is represented in the musical appearance of the chromosome. As an additional option, another genetic algorithm produces the mentioned optimal values and weightings, and by doing this allows the generation of rhythmic "families," groups of structures that differ in the occurrence of rhythmic characteristics. This Meta-GA facilitates evaluation by a user because the selection of particular rhythmic "families" leads more quickly to a rhythmic structure that is considered satisfying. Another function enables an efficient generation of rhythmic patterns by further structuring chromosomes regarding some parameters. The rhythmic structure of each chromosome is produced by different percussion instruments whose sound characteristics are distinguished by their respective parameter ranges from each other. Crossover and mutation are also only applied within similar instruments so that sudden changes of sound color may be avoided by genetic operations.

Other interesting approaches in which genetic algorithms are applied in the generation of rhythmic structure can be found in the works by Alejandro Pazos, A. Santos del Riego, Julian Dorado and J. J. Romero-Cardalda, [30, 31][13] who used artificial neural networks as fitness raters in their works.

Pazos et al. developed a model of a genetic algorithm in which two interacting populations of "artificial musicians" create different rhythmic structures that are evaluated by a user.

For the generation of his genetic algorithms, Burton used ART (short for Adaptive Resonance Theory) networks as fitness evaluators which are a type of network applied to the forming of categories within unordered data (see chapter 9). For the fitness evaluation, the network is trained with a corpus of drum patterns of different styles; the outputs of the genetic algorithm are compared by means of the resulting categories (clusters). An interesting aspect of this comprehensive work is also shown by a number of variants of genetic operations that are examined regarding their suitability for producing the rhythmic patterns.

In his work, Pearce treats the generation of rhythmic patterns in the style "drum and bass" and uses a multi-layered perceptron (see chapter 9) as a fitness function.

[13] See also the comprehensive developments in the dissertations of Anthony Richard Burton [6] and Marcus Pearce [32].

He legitimately criticizes the mostly inexact rating of the results of systems of algorithmic composition and counters this lack in his work with a number of evaluations: "In the present study an attempt has been made to evaluate in more objective terms whether the system fulfils the specified aims. In fact, aesthetic judgment has been removed from the evaluation of the generated patterns. This chapter describes three experiments: The first was a musical Turing Test of whether subjects could discriminate human from system generated patterns (section 5.2); the second asked subjects to classify the patterns according to style (section 5.3); and the final experiment asked for judgments of the diversity present in groups of three system generated patterns taken from both between and within runs (section 5.4). In a less objective manner a musician was asked to evaluate the system (section 5.5) and finally, the system was informally judged on its usability and practicality [. . .]" [32, p. 61]. The results of the evaluation show that patterns generated by humans were rated as better – a fact that, however, is caused by the occasionally unreliable fitness evaluation done by artificial neural networks rather than by the genetic algorithm itself.

7.4.5 Applications of Genetic Programming

In numerous applications of genetic programming, Lee Spector and Adam Alpern developed strategies for the automatic production of programs for the generation of musical structure. In a first approach [39], four-bar melodies serve as input generating melodies of the same duration as an output. The notes are represented as vectors of duration and pitch, each bar being given 48 of these pairs. This division allows for rhythmic variations of the melodic material ranging from full notes to 1/32 triplets, the length of the symbol string staying the same for every chromosome. For the genetic programming, different functions such as crab, reduction or copying of a musical fragment are chosen that are applied to the melodies of the input. In another project [40], Spector worked with specialized control structures and functions that enable different strategies of copying and transposing. In a multistage process, the results are evaluated both by a neural network and by a rule-based system – an approach allowing a significantly better output of the system in this project. For the automatic generation of control structures, Lee Spector [41] set automatically generated macros in contrast to the automatically generated functions. In the programming language Lisp that is applied here, macros refer to structures generating program codes. The advantages of a macro over a function can be seen e.g. in program sequences in which side effects play an important role. In Lisp, a side effect is a value assignment for a variable in the course of a function call. This effect may be desired explicitly; in the formulation of a control structure by means of a common function definition, however, undesired value assignments may result. If, for example, a control structure of the kind "if then else" is to be formulated in Lisp, then the following problem occurs in the definition as a function: In a common function definition in Lisp (*defun* name (parameter) (function body)) the parameters are analyzed before transferring them to the function body. So, a function definition of the kind *(defun if (if then else)* (function body)) will first carry out the value assignments of the forms "then" and "else" before evaluating the function body. Independent from the value of the "if" form, both following expressions are evaluated even though this is undesired. If, however, this function is defined as a macro, it behaves, if called, like an assignment of the form "if then else" common in other programming languages: Depending on the value of "if," either the "then" form or the "else" form is evaluated.

Brad Johanson described his method of applying genetic programming in the conception of his GPMusic system in [22], with Ricardo Poli [23]. GPMusic enables the production of melodies of variable length that, in the first version of the program, are rated by a user. The musical units used are the chromatic tones in a range of six octaves, a symbol for pauses, and arpeggiated chords from the tones of the C major scale (so-called "pseudo chords"). For genetic operations, the following functions are developed:

1. Play_two: This function takes two note sequences and concatenates them together.
2. Add_space: A rest is inserted after each time slot in the original sequence.
3. Play_twice: Repetition of the sequence.
4. Shift_up: Transposition one step up within the used scale.
5. Shift_down: Shift_up vice versa.
6. Mirror: The argument sequence is reversed.
7. Play_and_mirror: The argument sequence is reversed and concatenated onto itself.

In this approach, a number of productions have disadvantageous effects that either generate very short or very long monotone sequences. Significant improvements can be achieved, for example, by reducing the tone material to diatonic scale degrees as well as by automatically eliminating very short sequences. In the fitness rating done by a user, an interface allows for modifications such as repeating a sequence or setting its length. Further, a neural network is used for the fitness evaluation that may be able to avoid the restrictions of the fitness bottleneck, but lets melodies of inferior quality pass as adequate.

The fact that genetic programming may be applied efficiently, as shown in the example of the multiplexer, yet nevertheless faces significant problems in the field of music, can be explained by the requirements that the particular fitness functions have to meet. The operation mode of a multiplexer itself represents, at the same time, a perfect fitness function. In the musical domain the way to design the respective genetic algorithm or the genetic programming strategies has great influence on the processing and representation of the musical data. The decisive factor, however, remains the efficiency of the fitness function that in all described variants must evaluate the musical output in a conclusive way. If the criteria for "quality" are to be formulated easily, the sense of genetic generation may be questioned; if this is not the case, the difficulty lies in coherently evaluating musical information in a multitude of mutually influential factors. The abovementioned restrictions, however, occur mainly in the field of style imitation; if genetic algorithms are used for genuine compositional tasks, the application of these algorithms can yield interesting results. So, also in approaches that react interactively to the input of a user during processing, genetic algorithms may hold great potential. These approaches range from systems that, like GenJam, are located within a particular genre, to software implementations like Vox Populi that base their outputs on the comparison of partial structures of distinct tones. Some of the other applications described in the following section utilize the basic principles of genetic algorithms in an inventive way by simulating reproduction and interaction of musical units within artificial habitats.

7.4.6 Interactive Systems

In numerous works (see [1], [2], [3], [4], [5]), Al Biles described the functioning of his program GenJam, developed to improvise jazz music based on genetic algorithms. On the basis of given harmonic and rhythmic structures, GenJam generates melodies that are rated by a user. The starting point for the melodies to be produced is information about tempo, rhythmic articulation, parts to be repeated and chord progressions. The chromosomes are represented as binary strings within two populations indicating bars and phrases. A bar in the genetic population consists of the assigned fitness as well as other values that represent notes, ties and pauses. The chromosomes on the phase level dispose of pointers on each four bar units. The population size in GenJam comprises 48 phrases and 64 bars. The scale degrees for the single chromosomes are selected on the bar level with respect to a number of possible chord types, as can be seen in table 7.1.

Chord	Scale	Chord	Scale
CMaj7, C6, C	C D E G A B	C7#9	C Eb E G A Bb
C7, C9, C13	C D E G A Bb	C7b9	C Db E F G Bb
Cm7, Cm9, Cm11	C D Eb F G Bb	CmMaj7	C D Eb F G A B
Cm7b5	C Eb F Gb Ab Bb	Cm6	C D Eb F G A
Cdim	C D Eb F Gb G# A B	Cm7b9	C Db Eb F G A Bb
C#5	C D E F# G# A B	CMaj7#11	C D E F# G A B
C7#5	C D E F# G# A#	C7sus	C D E F G A Bb
C7#11	C D E F# G A Bb	CMaj7sus	C D E F G A B
C7alt	C Db D# E Gb G# Bb	Â	Â

Table 7.1 Allowed tone pitches regarding chord type in GenJam.

The genetic operation carried out on both chromosome types is a one-point crossover, one half of the symbol string resting unmodified, the other half being subject to a number of mutations (table 7.2). Holds are represented here by "15," pauses by "0." Four chromosomes of the bar level are chosen at random, the two with the highest fitness being selected for mutation; their production replaces the two other chromosomes. This population forms the basis for the structuring of the phrases that in turn may be submitted to several mutations. The operator "Invert" additionally changes pauses to holds and vice versa. "Genetic repair" randomly chooses a new order for the elements and also replaces the chromosome with the worst fitness by a randomly selected element. "Super phrase" generates a completely new phrase by using bars of greatest fitness selected from four groups that consist of three consecutive bars each. "Lick thinner" substitutes randomly selected bars that are chosen most frequently in order to avoid the generation of material that only differs slightly from an "optimal" solution. "Orphan phrase" works similar to "super phrase" with the difference that it selects those bars that occur least frequently. By means of these genetic mutations operating on the bar and phrase level, structurally similar phrases may be generated whose genetic variety is guaranteed by functions that work against the production of monotonous musical material. The output of the system is rated binary by the user during playback. The fitness obtained through this is assigned to each particular combination of phrases and bars and serves as a guideline for the further generation of populations. In an extension of the software, the

Mutation Operator	Mutated Measure
None (Original Measure)	9 7 0 5 7 15 15 0
Reverse	0 15 15 7 5 0 7 9
Rotate Right (e.g. 3)	15 15 0 9 7 0 5 7
Invert (15-value)	6 8 15 10 8 0 0 15
Sort Notes Ascending	5 7 0 7 9 15 15 0
Sort Notes Descending	9 7 0 7 5 15 15 0
Transpose Notes (e.g. +3)	12 10 0 8 10 15 15 0

Mutation Operator	Mutated Phrase
None (Original Phrase)	57 57 11 38
Reverse	38 11 57 57
Rotate Right (e.g. 3)	57 11 38 57
Genetic Repair	57 57 11 29
Super Phrase	41 16 57 62
Lick Thinner	31 57 11 38
Orphan Phrase	17 59 43 22

Table 7.2 Genetic mutation in bars (top) and phrases (below) in GenJam.

originally random selection of note values is controlled by algorithms that initialize the genetic populations by a structure similar to a reference corpus.

Another improvement is the integration of user improvisations that are used as a basis for the genetic mutations. Figure 7.14 shows the comparison of an improvisational input with the corresponding output of GenJam.

In a further experiment, Al Biles, Peter G. Anderson and Laura W. Loggi [5] implemented a neural network for the rating function. The neural network is trained

Fig. 7.14 Input of a user (top) and reaction of GenJam (bottom).

with statistical analyses of improvisations and serves to discard material that is musically unsuitable in advance of the genetic operations. However, the results do not meet expectations and show a clear advantage of the fitness evaluation carried out by a human expert.

Vox Populi by Jonatas Manzolli, Artemis Moroni, Ferdinando von Zuben and Ricardo Gudwin[14] is an interactive system that serves to generate musical structure in real time by means of genetic algorithms. In this approach, chromosomes are chords of four notes that are represented by binary symbol strings. During the output, the automatic fitness rating is evolved; in addition, during playback Vox Populi reacts to user inputs that have an influence on the different fitness functions as well as on the temporal processing of the genetic algorithm. At the beginning of the genetic cycle, a population of thirty chords is generated whose notes are assigned randomly selected MIDI values between 0 and 127. Consequently, each of these chromosomes has different pitch information for each chord tone; all the other musical parameters such as rhythmic structure or timbre are given by the user and may also be modified during playback. The genetic operations are crossover and mutation processes typically applied in binary symbol strings. Fitness is evaluated with regard to voice ranges and consonance relations within the individual chords.

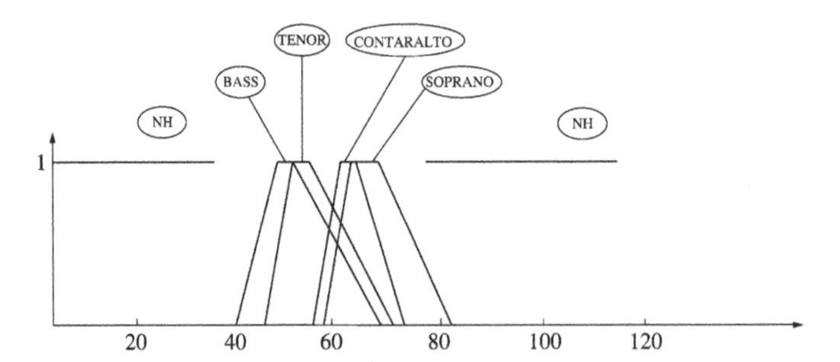

Fig. 7.15 Initial values for voice ranges in the fitness evaluation in Vox Populi. [25, p. 5]. With kind permission of the authors.

For the voice range fitness, rough areas of four distinct voices (soprano, tenor, bass, alto) are assumed as fuzzy sets, as shown in figure 7.15. Depending on how

[14] [25], without Gudwin in [28].

many of the chord voices lie within the fuzzy set, the voice ranges are assigned a fitness rating between 0 and 1. The fitness rating for harmonic consonance compares chord notes in regard to corresponding frequencies in the upper harmonic tones. This fitness evaluation takes place between the notes of a chord and also between a given note (tonal center) and the chord notes. The interface of "Vox Populi" allows for the modification of the following parameters during the playback of generated musical structure: The "tonal center control" enables the modification of the tonal center. "Biological control" determines the slice of time necessary to apply the genetic operation. "Rhythmic control" indicates the slice of time necessary to evaluate the fitness and consequently also to generate a new population. "Voice range control" allows enlarging or diminishing the voice ranges represented in figure 7.15. "Orchestra control" assigns different instruments to the musical structures.[15] Another input window enables the drawing of curves that are linked to each of the two abovementioned parameters. "Performance control" enables the adjustment of settings related to differing temporal performance of harmonic constellations as well as the selection of particular scales that serve as a basis for the genetic operations; again, these scales naturally depend on the currently selected tonal center.

Sidney Fels and Jonatas Manzolli [10] developed an interactive system for algorithmic composition that is controlled by the movements of two users in a space. The outputs of the system are generated by the interaction of nested modules as shown in figure 7.16.

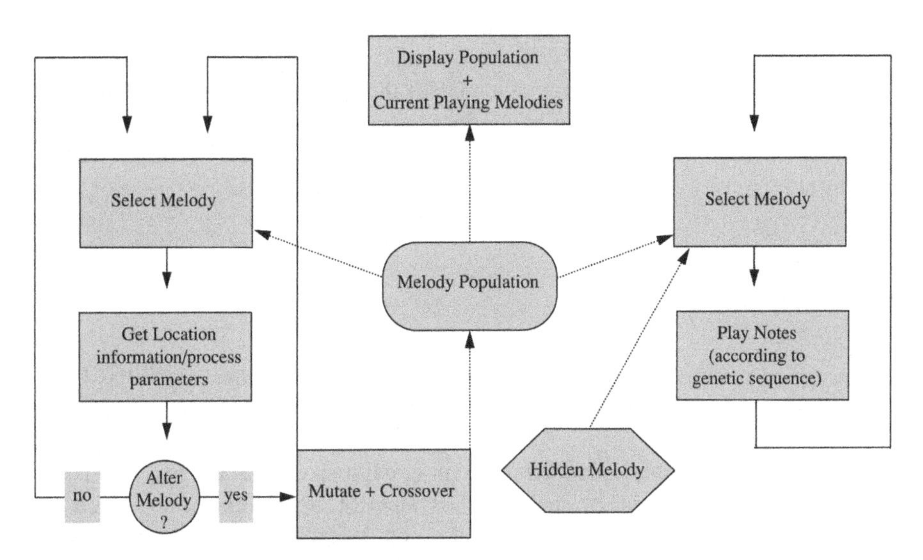

Fig. 7.16 Components of the system of Fels and Manzolli.

[15] In this case: Strings, brass instruments, keyboards and percussion, as well as a random orchestra that takes any instrument from the General-Midi list.

The system is initialized with melodies in a range of two octaves; every tone pitch is assigned a duration and a velocity. These melodies each make up a population of twenty chromosomes that are partly subject to crossover and mutation. Within the "Musical Cycle," a number of chromosomes are selected by means of pattern comparison with a "hidden melody."[16] This reference melody is established a priori, does not affect the evolution of the population and also is not subject to any genetic operations. It only determines which melody of the current population is played. This cycle repeats at a particular rate, allowing a polyphonic overlapping of the generated melodies in higher values. The genetic operations of crossover and mutation are carried out in the "Biological Cycle" and affect velocity, pitch and duration. The mutation modifies musical parameters in fragments of a chromosome and crossover generates a new chromosome by swapping fragments of two randomly selected "parent chromosomes." The degree of application of mutation and crossover to the chromosomes of a population is determined by weighted probabilities that result from the position of two objects – in a later version two users – in a space. The movements of the objects or users are captured by a tracking system. The physical proximity of the users is directly proportional to the mutation probability and the relative angle between the users and a video wall is indirectly proportional to the crossover probability. On the wall, the melodies are represented as zigzag lines whose lengths give the durations and their angles the tone pitches. An inventive view of a biological system is shown by the fact that the "shared line of sight" of the users on the video wall is awarded with intensive crossover behavior and "distance" is punished with "mutated" offspring.

An interesting aspect in the work of Fels and Manzolli is the absence of any kind of fitness function. This approach enables the system to produce a constantly changing output that, on the one hand, does not approximate an "ideal state," but on the other hand, moves within a musical range which is given by the respective structural proximity of the chromosomes to the "hidden melody."

7.4.7 Artificial Life-Forms

By means of genetic algorithms, Peter Todd and Gregory Werner, as well as Palle Dahlstedt and Matts Nordahl, simulated artificial life-forms in which "individuals" interact in different ways.

Todd and Werner [44] examined different conditions for passing over and evaluation of melodic material by means of a system of algorithmic composition. A number of "males" are selected by "females" on the basis of the attractiveness of their "love songs" (32 tone pitches from a two-octave range) for mating that produces offspring through genetic operations. Each female disposes of a matrix of transition probabilities representing the expectations for particular transitions. The matrices of the female population are initialized with simple folk song melodies and

[16] Cf. also: target pattern in [18], pattern comparison in [11], [12], and [13]; the "hidden melody," however, does not act as a fitness function in contrast to these approaches.

compared with the productions of the males in three different ways. In the first case, each new transition that is "heard" is compared with the transition probabilities of the matrix and all obtained values are finally added to the evaluation of the "love song." If, for example, the minor sixth scored highest in the comparison matrix – meaning that it occurred most frequently in the coded songs – a "love song" by a male consisting mainly of this transition receives very high ratings accordingly. Because only the attractiveness of every single interval step is measured in this so-called "local transition preference," an evaluation of the entire musical structure is therefore deliberately avoided. In the second method, the "global transition preference," a matrix with transition probabilities is generated for each male production that is compared with the comparison matrix of the female. So, in this case, similarity is evaluated with regard to the total structure of the "love song." In order to avoid the development of unvaried material, the "surprise preference" is implemented as a last method. Here, an evaluation occurs based on the sum of each difference between the transition with the highest expectation and every single transition of the "love song," meaning that precisely the most surprising movements – the ones occurring least frequently in the comparison matrix – are rated highest. An additional input variable (2 to 20) determines how many selected males are evaluated by the females at the same time. Through a combination of the abovementioned evaluation possibilities, each female finally chooses a male for mating and the process is carried out by crossover or mutation of the two genomes. The offspring is copied into the population and when a critical amount is reached, a randomly chosen third of the population is deleted. Each female generates one offspring with a male partner. With the males, the evaluation of their melodies decides the number of mating partners. As a further extension of this model, Todd and Werner allow modifications of the female transition tables by coevolving with the productions of the males – an alternative that leads to greater variability in the gene pool.

Palle Dahlstedt and Matts Nordahl [7] developed an artificial world of interacting creatures that can walk, eat and search for mating partners. The individuals move on a grid and can walk in eight different directions; these are either limited by walls, or the grid has a toroidal structure.[17] The creatures move in time cycles that also influence the lifetime of the individuals. Chances of survival can be improved by eating the food stored on the grid. Activities such as singing, walking and mating reduce life points that are thrown out in the world as randomly distributed food. The genome of the individuals consists of two parts. The "sound genome" contains melodies of up to ten notes, the first note indicating the species to which the creature belongs. The "procedural genome" controls the actions of the individual as an associated function that is generated by genetic programming.[18] A creature dies if its life points decrease below a particular threshold, when a given age limit is exceeded or when the genome length exceeds a critical value. The criteria for mating result from the requirement for a minimal number of necessary life points, sexual maturity (expressed by a particular number of lived time steps) as well as the necessity

[17] For torus, see chapter 8.

[18] The instructions used here are: "WALK," "TURN," "REST," "SING," "LOOP" and "IF," whereas "LOOP" and "IF" cannot be nested and "TURN" may be used for different directions.

of a direct proximity to a mating partner of the same species, expressed by the same first note or a same first interval. A further criterion for successful reproduction that occurs by means of crossover of the single genomes is the existence of suitable melodies that do not necessarily have to come from the selected partner. Dahlstedt's and Nordahl's system features a number of other parameters for the production of musical structure which always makes the total activity of the "biological" system audible.

The works of Todd and Werner[19], as well as those of Dahlstedt and Nordahl, show interesting application possibilities of genetic algorithms in systems of algorithmic composition that refer to the behavior of individuals in biological habitats in an inventive way.

Another approach that deals with biological processes is evident in some works that make use of the sonification of genetic structure. In these, however, the search for mapping strategies for an adequate representation of genetic sequences in musical structure is in the foreground [15, 14, 9]. A number of systems apply genetic algorithms to sound synthesis, a field that is, however, not in the scope of this work.

7.5 Synopsis

Genetic algorithms as a method of heuristic state space search have, in addition to neural networks, a special position among the methods of algorithmic composition.[20] The architecture of this class of algorithms promotes the production of a large number of small form segments in the generation of musical structure. An essential feature of a genetic algorithm is the continuous generation and examination of symbol strings – a procedure that is highly suitable for process-like compositional concepts. Besides the fitness evaluation, which is of crucial importance, the way the musical information is encoded in the chromosomes plays an essential role. Among the different methods mentioned, the decision for an absolute or relative encoding of tone pitches has the greatest effect on the results of the genetic operations. When a chromosome symbol is modified, in an absolute representation (e.g. designating concrete tone pitches) the following pitches remain untouched. In a relative encoding (intervallic), all succeeding pitches change corresponding to the value of the transposition. The relative representation allows for tone pitch transposition of whole segments and by this, provides a traditional procedure of musical structure modification. If, however, the transposition only applies to one voice of a polyphonic movement or an undesired abrupt modulation occurs through this operation, an absolute representation of the chromosomes will be preferable. The mutation of a chromosome symbol produces a larger modification of absolute tone pitches in

[19] See also [42] – in this article, numerous other approaches of genetic algorithms are treated in detail, particularly considering the differently generated fitness functions. Further interesting reading should be found in the newly published book edited by Miranda and Biles [27].

[20] Markov models, generative grammars, transition networks, Lindenmayer systems and cellular automata are equivalent in some configurations.

relative representation as compared to the absolute variant. This makes the effects of the mutation, on the one hand, more difficult to foresee, on the other hand, a larger genetic variety is produced within one generation. Better results may generally be achieved in genetic algorithms with representation procedures that represent the complexity of different aspects of musical information. An interesting aspect besides the abovementioned strategies is given by Michael Mozer,[21] who develops a multi-dimensional representation system based on perceptive principles for his neural network.

The operations of the traditional genetic algorithm, such as crossover and mutation are, in some works, extended or replaced by manipulating the musical material by means of mirror, crab, etc. The application of this approach, however, is not always suitable because on the one hand, these musical transformations do not represent generally accepted principles, but are, for example, only used in the context of particular styles in order to create formal relations between musical components in the western musical tradition. On the other hand, another problem arises in the application of these transformations within a genetic algorithm. The transformations of crab, mirror and simple transpositions refer in general to musical segments formally belonging together. Because, however, a genetic algorithm generally generates new outputs by continuously arranging fragments of chromosomes, correspondingly, different transformations within a chromosome are combined and therefore lose their structural functions. A genetic algorithm achieves locally acceptable results very fast; the generation of larger musical fragments, however, is a major problem because, in analogy to the above example, in every genetic operation the context is completely changed within the chromosome. If, for example, the chromosomes represent chord progressions, it is highly probable that with every crossover an undesired abrupt modulation also results. As shown in the work of Mc Intyre, a large population and a large number of generations are therefore not able to respond to this problem adequately. If, according to this, specific knowledge about the domain to be modeled is available for the task of style imitation, rule-based procedures naturally prove to be superior to genetic algorithms as demonstrated very clearly in the examination of Phon-Amnuaisuk and Wiggins. Even though no domain-specific knowledge is available, but a corpus of correct compositions for the examination of style conformity may be provided, the application of other procedures such as Markov chains or generative grammars is recommended because these are better able to process context-dependent information. These techniques also have specialized methods that can better accomplish the generation of globally acceptable structure.[22]

The fitness evaluation in the genetic algorithm is the decisive factor with regard to the quality of the generated material. The fitness function may be carried out with different evaluators. If the evaluation is not performed by a human user, which implies the previously mentioned problem of fitness bottleneck, rule-based procedures or neural networks may be applied for the examination of the generated material. An

[21] Cf. Michael Mozer's work treated in chapter 9.

[22] E.g. the Viterbi algorithm in hidden Markov models (chapter "Markov Models") or Kohonen's "Self-Learning-Grammars" (chapter 4).

algorithmic fitness evaluation of this kind, however, has two serious disadvantages. First, the evaluated algorithm could itself undertake the task of generating material instead of only acting as an instance of examination – in this case, the genetic algorithm only provides a great number of material to be examined that rarely yields correct results due to the hardly feasible treatment of context in the generation process. Secondly, the creation of an algorithmic fitness evaluation – except for simple rules of music theory – is in a musical context a very difficult undertaking because the fitness function in this case would have to represent an absolutely valid quality criterion for the judgment of a number of mutually connected and influencing musical factors. This is, firstly, extremely difficult to put into practice and secondly, makes the genetic algorithm itself unnecessary in the framework of a possible implementation within a particular knowledge-based system. One principal alternative for the improvement of evaluations is to apply fitness functions to a series of different quality criteria, and also to numerous stages, as is done in some of the mentioned works. However, here too, the problem often arises whereby, in improving a certain fitness dimension, the already received fitness of another category is worsened again.

Therefore, for the field of style imitation, the development of an algorithmic fitness evaluation is, due to the aforementioned reasons, a challenging problem. On the other hand it is precisely the specific behavior of the genetic algorithm that makes it so suitable for a number of alternative tasks. The permanent modification of musical material, the generation of a great number of differently interpretable amounts of data, as well as the immanently time-dependent aspect – the permanent passing of the multiple cycle of production, modification and examination – are aspects of great interest for numerous compositional approaches. A further strength of the genetic algorithm is also shown in some of the previously mentioned systems that in an innovative way use behavior in a "biological habitat" as a source of inspiration.

References

1. Biles JA (1994) GenJam: a genetic algorithm for generating jazz solos. In: Proceedings of the 1994 International Computer Music Conference. International Computer Music Association, San Francisco
2. Biles JA (1995) GenJam Populi: training an IGA via audience-mediated performance. In: Proceedings of the 1995 International Computer Music Conference. International Computer Music Association, San Francisco
3. Biles JA (1998) Interactive GenJam: integrating real-time performance with a genetic algorithm. In: Procedings of the 1998 International Computer Music Conference. International Computer Music Association, San Francisco
4. Biles JA (1999) Life with GenJam: interacting with a musical IGA. In: Proceedings of the 1999 IEEE International Conference on Systems, Man and Cybernetics, vol 3. IEEE, Piscatory, NJ, pp 652–656
5. Biles JA, Anderson PG, Loggi LW (1996) A neural network fitness function for a musical IGA. In: Anderson PG, Warwick K (eds) IIA '96/SOCO '96: first international ICSC symposium on intelligent industrial automation (IIA '96) and soft computing (SOCO '96). ICSC, Millet, Alberta, pp B39-B44

6. Burton AR (1998) A hybrid neuro-genetic pattern evolution system applied to musical composition. Dissertation, University of Surrey, 1998
7. Dahlstedt P, Nordahl MG (2001) Living melodies: coevolution of sonic communication. Leonardo, 34/3, 2001
8. Darwin C (1859) On the origin of species by means of natural selection, or the preservation of favoured races in the struggle for life. John Murray, London
9. Dunn J, Clark MA (1999) The sonification of proteins. Leonardo, 32/1, 1999
10. Fels S, Manzolli J (2001) Interactive, evolutionary textured sound composition. In: Jorge JA, Correia NM, Jones H, Kamegai MB (eds) Multimedia 2001: Proceedings of the Eurographics workshop in Manchester, United Kingdom, September 8–9, 2001. Springer Wien, pp 153–164
11. Gartland-Jones A (2002) Can a genetic algorithm think like a composer? In: Proceedings of 5th International Conference on Generative Art, Politecnico di Milano University, Milan, 2002
12. Gartland-Jones A (2003) MusicBlox: a real-time algorithmic composition system incorporating a distributed interactive genetic algorithm. In: Raidl G et al (eds) Applications of evolutionary computing. Lecture notes in computer science, vol 2611. Springer, Berlin, pp 490–501
13. Gartland-Jones A, Copley P (2003) The suitability of genetic algorithms for musical composition. Contemporary Music Review, 22/3, 2003
14. Gena P, Strom C (2001) A physiological approach to DNA music. In: Proceedings of CADE 2001: the 4th Computers in Art and Design Education Conference, Glasgow, 2001
15. Gena P, Strom C (1995) Musical synthesis of DNA sequences. XI Colloquio di Informatica Musicale, Univeristà di Bologna, 1995
16. Goldberg DE (1989) Genetic algorithms in search, optimization, and machine learning. Addison Wesley, Bonn. ISBN 0-201-15767-5
17. Holland J (1975) Adaption in natural and artificial systems. University of Michigan Press, Ann Arbor, Michigan
18. Horner A, Goldberg DE (1991) Genetic algorithms and computer-assisted music composition. In: Proceedings of the 1991 International Computer Music Conference. International Computer Music Association, San Francisco
19. Horowitz D (1994) Generating rhythms with genetic algorithms. In: Procedings of the 1994 International Computer Music Conference. International Computer Music Association, San Francisco
20. Jacob BL (1994) Composing with genetic algorithms. In: Proceedings of the 1994 International Computer Music Conference. International Computer Music Association, San Francisco
21. Jacob BL (1996) Algorithmic composition as a model of creativity. Organised Sound, 1/3, December, 1996
22. Johanson B (1997) Automated fitness raters for the GPMusic system. Masters Degree Final Project, University of Birmingham, 1997
23. Johanson B, Poli R (1998) GP-Music: an interactive genetic programming system for music generation with automated fitness raters. In: Koza JR et al (eds) Genetic programming 1998. Morgan Kaufmann, San Francisco
24. Koza JR (1992) Genetic programming: on the programming of computers by means of natural selection. MIT Press, Cambridge, Mass. ISBN 0-262-11170-5
25. Manzolli JA, Moroni F, Von Zuben R, Gudwin R (1999) An evolutionary approach applied to algorithmic composition. In: Proceedings of SBC99 - XIX National Congress of the Computation Brazilian Society, Rio de Janeiro, 3, 1999
26. Mc Intyre RA (1994) Bach in a Box: the evolution of four part Baroque harmony using the genetic algorithm. In: Proceedings of the First IEEE Conference on Evolutionary Computation, Orlando, Florida, 1994, vol 2. IEEE, Piscataway, NJ, pp 852–857
27. Miranda ER, Biles JA (eds) (2007) Evolutionary computer music. Springer, London. ISBN 13 978-1-84628-599-8
28. Moroni A, Manzolli J, Von Zuben F (2000) Composing with interactive genetic algorithms. In: Proceedings of SBC2000 – Congresso da Sociedade Brasileira de Computação, 2000

29. Papadopoulos G, Wiggins G (1998) A genetic algorithm for the generation of jazz melodies. STeP98, Jyväskylä, Finland, 1998
30. Pazos A, Santos A, Dorado J, Romero-Cardalda JJ (1999) Adaptative aspects of rhythmic composition: Genetic Music. In: Banzhaf W (ed) GECCO '99: proceedings of the Genetic and Evolutionary Computation Conference, vol 2. Morgan Kaufmann, San Francisco
31. Pazos A, Romero-Cardalda JJ, Santos A, Dorado J (1999) Genetic Music Compositor. In: Proceedings of 1999 Congress on Evolutionary Computation, vol 2. IEEE, Piscataway, NJ. pp 885–890
32. Pearce M (2000) Generating rhythmic patterns: a combined neural and evolutionary approach. Dissertation, Department of Artificial Intelligence, University of Edinburgh, 2000
33. Phon-Amnuaisuk S, Tuson A, Wiggins G (1999) Evolving musical harmonisation. In: Dobnikar A, Steele NC, Pearson DW Albrecht RF (eds) Artificial neural nets and genetic algorithms: proceedings of the International Conference in Portoroz, Slovenia, 1999. Springer, Wien, pp 229–234
34. Phon-Amnuaisuk S, Wiggins GA (1999) The four-part harmonisation problem: a comparison between genetic algorithms and a rule-based system. In: Proceedings of the AISB '99 Symposium on Musical Creativity. Society for the Study of Artificial Intelligence and Simulation of Behaviour, Edinburgh
35. Pigg P (2002) Cohesive music generation with genetic algorithms. http://web.umr.edu/ tauritzd/courses/cs401/fs2002/project/Pigg.pdf Cited 11 Nov 2004
36. Rechenberg I (1973) Evolutionsstrategie – Optimierung technischer Systeme nach Prinzipien der biologischen Evolution. Frommann-Holzboog, Stuttgart
37. Santos A, Arcay B, Dorado J, Romero J, Rodrìguez J (2000) Evolutionary computation systems for musical composition. In: Mastorakis NE (ed) Mathematics and computers in modern science. World Scientific and Engineering Society Press, Athens. ISBN 960 8052 238
38. Seebach G (2004) Evolutionsstrategien. http://www.seebachs.de/Evolutionsstrategien.html Cited 4 Mar 2005
39. Spector L, Alpern A (1994) Criticism, culture, and the automatic generation of artworks. In: Proceedings of the Twelfth National Conference on Artificial Intelligence. American Association for Artificial Intelligence, Menlo Park, Calif
40. Spector L, Alpern A (1995) Induction and recapitulation of deep musical structure. In: Working Notes of the IJCAI-95 Workshop on Artificial Intelligence and Music
41. Spector L (1995) Evolving control structures with automatically defined macros. In: Working Notes of the AAAI Fall Symposium on Genetic Programming. American Association for Artificial Intelligence, Menlo Park, Calif
42. Todd PM, Werner GM (1998) Frankensteinian methods for evolutionary music composition. In: Griffith N, Todd PM (eds) Musical networks: parallel distributed perception and performance. MIT Press/Bradford Books, Cambridge, Mass. ISBN 0262071819
43. Towsey M, Brown A, Wright S, Diederich J (2001) Towards melodic extension using genetic algorithms. Education Technology & Society, 4/2, 2001
44. Werner GM, Todd PM (1997) Too many love songs: sexual selection and the evolution of communication. In: Husbands P, Harvey I (eds) (1997) Fourth European Conference on Artificial Life. MIT Press/Bradford Books, Cambridge, Mass. ISBN 0262581574
45. Wiggins G, Papadopoulos G, Phon-Amnuaisuk S, Tuson A (1999) Evolutionary methods for musical composition. International Journal of Computing Anticipatory Systems, 1999

Chapter 8
Cellular Automata

Cellular Automata (*CA*) are, for the most part, used to model discrete dynamic systems. The temporal development of the system is represented in an n-dimensional cell space. The cell space is represented by vectors or n-dimensional matrices whereas the dimension is assumed to be principally infinite. Within this space there are cells that may assume a finite number of states. The state of a cell at a discrete timestep t_0 is determined by its own state as well as the states of the neighboring cells at timestep t_{-1}. In nearly every type of cellular automaton, only discrete changes of state are taken into consideration. Each cell of the grid must follow the same state transition rules. A new state of a cellular automaton results from the application of the state transition rules to all cells of the cell space.

8.1 Historical Framework and Theoretical Basics

From 1922 on, the German physicist Wilhelm Lenz and his student Ernst Ising examined the properties of ferromagnetism. The so-called "Ising model" acts based on the assumption that, in magnetizing a ferromagnetic substance, the magnetic moments of the atoms can only assume two states. Furthermore, paired interactions between neighboring particles occur that eventually lead to a state of collective order in the system. The "Ising model" may be considered as a general model of interacting particles and may also be applied to phenomena such as the condensation in gases. Because the interactions in the "Ising model" occur in analogy to the rules in a cellular automaton, this model may be considered a forerunner of this class of algorithms.

The first theories concerning cellular automata in the strictest sense were devised in the 1950s by Konrad Zuse, Stanislav Marcin Ulam and John von Neumann. Konrad Zuse developed an alternative description of physics with a model he calls the "Calculating Space" [34]. In this virtual space, information is exchanged by means of interactions of adjacent cells. The elementary units of this "information continuum" are represented by different states that the cells can assume. Beginning in

Fig. 8.1 Stanislav Martin Ulam. Courtesy Los Alamos National Laboratory. Reproduced with kind permission.

1942, when Ulam was working under the direction of Oppenheimer on the "Manhattan project," he developed in his free time a principle for the computer in Los Alamos that allows for creating complex graphic patterns by means of interacting objects to be produced by a printer. His "recursively defined geometric objects" are simply cells of a cellular automaton whose state changes occur based on the states of the circumjacent cells. Inspired by Ulam's concepts, von Neumann developed groundbreaking ideas for the application of this newly discovered principle amongst others by using the model of a universal computer with the ability to self-reproduce. Due to his early death in 1957, von Neumann was unable to finish this line of research. His considerations were later further developed by the American philosopher and computer scientist Arthur W. Burks, who also published von Neumann's "Theory of Self-Reproducing Automata" [28] in 1966. In the beginning of the 1970s, cellular automata gained unexpected popularity due to John Horton Conway's "Game of Life" (GoL), which was presented to a broad public in the monthly rubric on mathematical games of the magazine "Scientific American" [15]. Conway worked with different initial configurations of his cellular automaton for a few years in order to evoke different patterns of behavior. The behavior of cellular automata became the subject of broader scientific examination. William Gosper, an expert on artificial intelligence at the Massachusetts Institute of Technology (MIT), resolved one of Conway's integral issues concerning self-reproducing structures within the GoL with his "glider gun." Furthermore, cellular automata also find application in the simulation of physical processes. Ed Fredkin founded the Information Mechanics Group at the MIT that was increasingly devoted to this aspect of cellular automata. In 1987, Fredkin's colleagues Tomaso Toffoli and Norman Margolus published a study [31] which dealt with the modeling of physical processes as well as basic theoretical reflections on the application of cellular automata. Carter Bays of the University of South Carolina worked to extend cellular automata into the third dimension, modeling his work on the "Game of Life." [1].

Fig. 8.2 Stephen Wolfram. © 2002 Stephen Wolfram, LLC. All rights reserved. Used with permission.

The British mathematician Stephen Wolfram considers the theory of cellular automata an evolution of the whole scientific field. In the preface of his book "A New Kind of Science," published in 2002, he writes: "Three centuries ago science was transformed by the dramatic new idea that rules based on mathematical equations could be used to describe the natural world. My purpose in this book is to initiate another such transformation, and to introduce a new kind of science that can be embodied in simple computer programs." [33, p. 1]. According to Wolfram, cellular automata make revolutionary models and problem solving strategies possible for a wide field of natural scientific and humanistic as well as artistic disciplines.

8.2 Types of Cellular Automata

8.2.1 1-Dimensional Cellular Automata

The simplest form of a CA consists of cells that may assume two states dependant on two neighboring cells. The state of the adjacent cell and that of the given cell make up the so-called "neighborhood" which may assume 2^3 possible configurations. From the eight possible states of the neighborhood that each may cross over into two possible states of the given cell, 2^8 possible transition descriptions for this class of CA result. Naturally, more distant cells whose number is determined by the "radius" (R) may also be involved. So, given $R = 2$, for example, the four adjacent cells, as well as the given cell are considered, enabling 2^5 possible configurations of the neighborhood. The neighborhood and the states in a 1-dimensional cellular automaton are also sometimes represented by $K(m)R(n)$, $K(m)$ indicating the number of cell states and $R(n)$ representing the radius of the neighborhood. The states of a neighborhood are usually noted as binary symbol strings in a row. In the next row, the resulting successive states of the given cells are also represented as binary values. The value of the second row specifies the number of the transition description or the rule of the CA, always assuming the same order of the state groups in the first row (configurations of three, arranged as binary values, ascending from right to left (see figures 8.4 and 8.5). Wolfram classifies 1-dimensional cellular automata into four basic types:

- Class 1: The CA shows simple behavior and evolves to a homogeneous final state.
- Class 2: The CA leads to different final states consisting of simple stable or periodically repeating structures.
- Class 3: The CA shows apparently chaotic and random behavior.
- Class 4: The CA produces complex patterns that may also repeat self-similarly.

A graphic representation of the behavior of these classes is indicated in figure 8.3. In all images, one single state is shown as one line; evolution in time begins at the top line and moves downward. The depiction of the behavior of the automata within these classes is simplified here to a large extent. Moreover, it must be considered that a CA on a finite toroidal (see below) grid will always display periodic behavior due to the fact that the possibilities with regards to placing cells on this grid are finite. So, if a CA perpetually generates new settings of cells (patterns) on a grid, a pattern must inevitably recur at some point. Because a traditional CA is based on deterministic rules, from now on, the patterns on the grid will constantly be repeated within this period. Given an infinite grid and a correspondingly infinite memory in a computer, a particular CA could naturally and continually generate its patterns anew without ever lapsing into a cycle.[1]

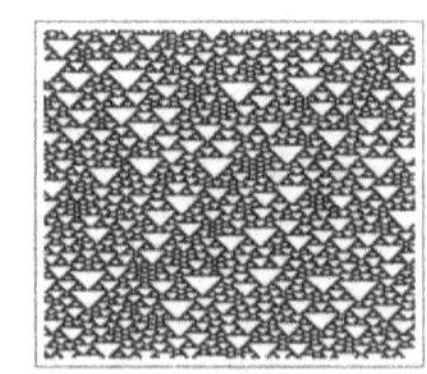

Fig. 8.3 The four classes (from left to right) of 1-dimensional cellular automata [33, p. 231]. © 2002 Stephen Wolfram, LLC. All rights reserved. Used with permission.

Figure 8.4 shows a cellular automaton of class 3 of rule 30. The representation depicts the changes of the CA in a cycle from top to bottom.[2]

[1] But due to the fact that even space is curved, this innovative CA will, unfortunately, also fall into a periodical cycle.

[2] The graphic representations were produced with the software program Mcel [32], unless otherwise noted.

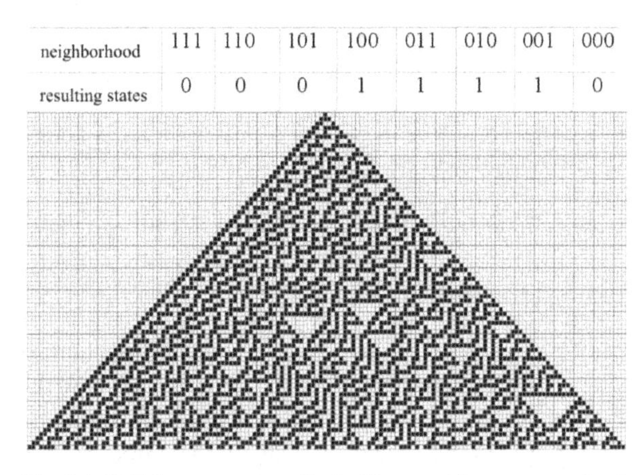

neighborhood	111	110	101	100	011	010	001	000
resulting states	0	0	0	1	1	1	1	0

Fig. 8.4 Cellular automaton of rule 30 – graphic representation in 75 iteration steps.

Figure 8.5 shows a cellular automaton of class 4 of rule 90.

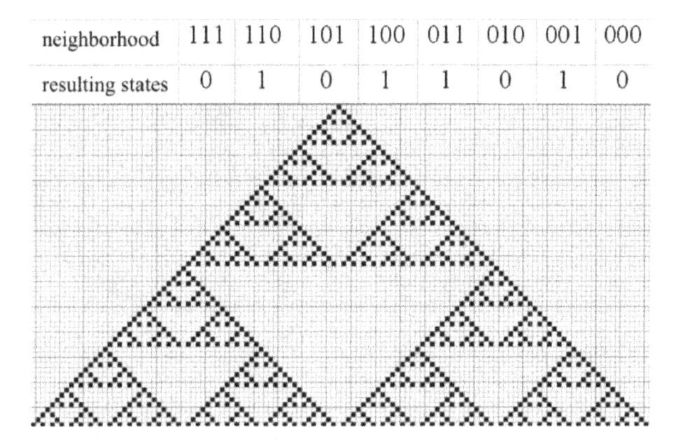

neighborhood	111	110	101	100	011	010	001	000
resulting states	0	1	0	1	1	0	1	0

Fig. 8.5 Cellular automaton of rule 90 – graphic representation in 63 iteration steps.

The structure resulting from this CA represents a self-similar so-called "Sierpinski triangle"; self-similar structures may also be easily generated with a Lindenmayer system (see chapter 6).

8.2.2 2-Dimensional Cellular Automata

The best known example of a 2-dimensional cellular automaton is Conway's "Game of Life." The initial configuration is presented as follows: "To play life you must have a fairly large checkerboard and a plentiful supply of flat counters of two colors. (Small checkers or poker chips do nicely.) An Oriental go board can be used if you can find flat counters that are small enough to fit within its cells." [15, p. 1]. An arbitrary arrangement of counters forms the initial point. This CA follows three basic rules:

- If an empty cell has exactly three adjacent neighbors, a counter is placed on it at the next move.
- Every counter with two or three neighboring counters survives for the next generation.
- In every other case, no counter is placed on the cell.

The term "checkerboard" must be seen as an analogy here, because in contrast to board games, the evolving patterns are generated by the state changes of the individual cells of the system, and not by changing the local positions of single components. The dimension of a CA's grid is generally assumed to be infinite. In a finite observation space, the grid behaves similarly to a torus (figure 8.6). Because the surface of a torus may be depicted without singularities on a square area, in a 2-dimensional representation, "leaving" the observed grid on one side leads to "entering" it on the opposite side.

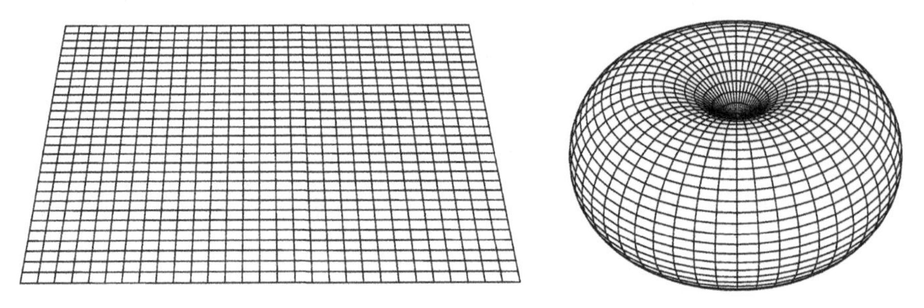

Fig. 8.6 2-dimensional grid mapped on a torus.

In figure 8.7, a simple configuration forms the basic pattern of the CA. By applying the rules of the "Game of Life," in this initial configuration a development occurs that finally grinds to a halt by oscillating between two stationary patterns.

Fig. 8.7 Variation of a simple initial configuration by the rules of the GoL.

The complex behavior of a cellular automaton in the context of the "Game of Life" shows particularly in a widely spread initial configuration. In some initial configurations of the GoL, some patterns have the ability to "move" over the grid of the CA without changing structure. Others change their initial configuration and are recreated after a certain number of cycles.

For the description of these and similar phenomena in particular initial configurations, a specific terminology is used. For example, "space ships" refer to finite patterns that reappear after a number of generations in changed position but with same orientation. "Puffers" move over the grid emitting elements of higher ("clean puffer") or lower ("dirty puffer") order.[3]

8.2.3 3-Dimensional Cellular Automata

If a 2-dimensional CA is represented within a 3-dimensional structure, where the third axis indicates the steps of the temporal development, this leads to a pseudo-3-dimensional CA, as shown in figure 8.8.

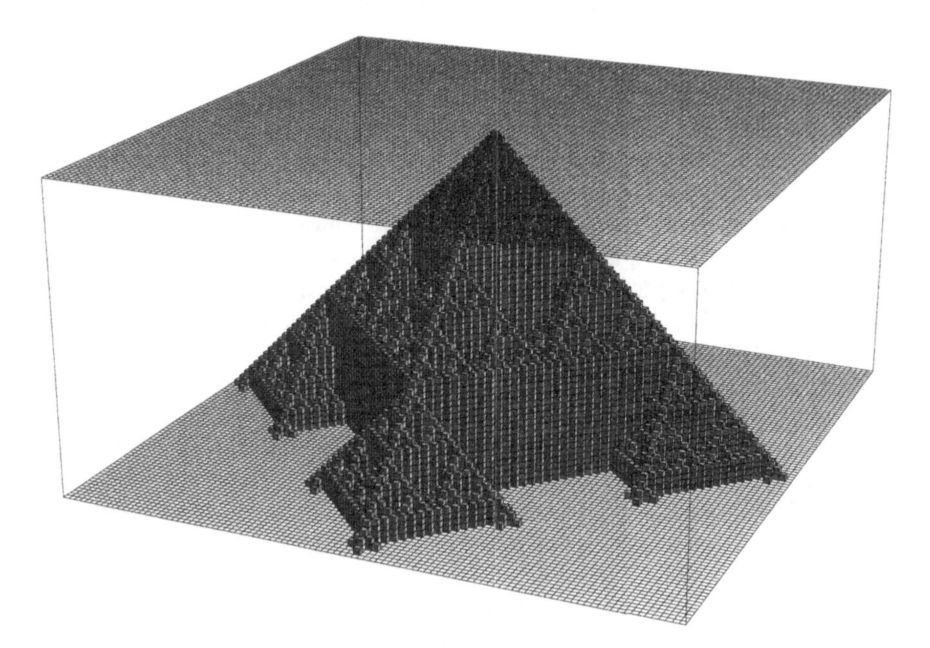

Fig. 8.8 Pseudo-3-dimensional CA [33, p. 172].

The chronological development of a real 3-dimensional CA may be represented by means of a cube in which, however, 26 cells must always be taken into account. Figure 8.9 illustrates some of the stages of a 3-dimensional CA with a very simple generation rule: A cell is activated when precisely one of its adjacent cells is active, whereas already active cells remain in their active state.

[3] For a detailed description of the different types see [30].

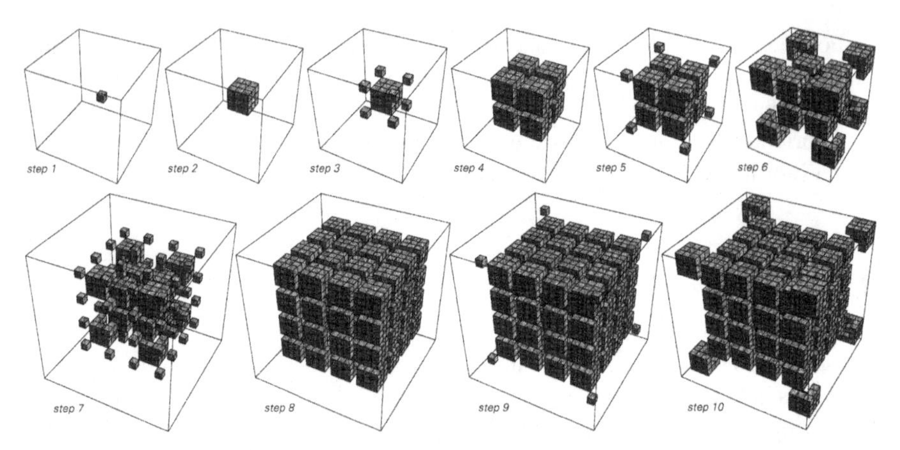

Fig. 8.9 Chronological development of a 3-dimensional CA [33, p. 183].

8.2.4 Extended Types of Cellular Automata

Whereas the state changes of a cellular automaton generally occur in discrete steps, functions that provide continuous values, usually between 0 and 1, may also be used for these processes; these automata are commonly referred to as *continuous automata* [33, p. 921–922]. Another extension of the traditional CA is given by the implication of chance. By means of a probability-based selection, different rules for the same cell configuration may be applied in this case. As another option, the rule of the CA may also define a probability for the next state of the cell [33, p. 922, 591–592]. In the types of automata mentioned so far, a rectangular grid is assumed; other shapes, however, may also be chosen.[4] A cellular automaton may also be created in more than three dimensions.[5]

These variants are only a few examples of extended formalisms of cellular automata: "In a general way, it is possible to build any type of automata by playing on structural and functional rules. The first ones define the spatial structure of the automata network, that is its number of dimensions, the disposition of cells (squares, hexagons,... in a two dimensional automaton) and the type of neighborhood determination. The second ones will determine the number of states and the transition rules." [29, p. 5].

[4] E.g. CA with triangular cells and 6 neighbors each, cf. [2].

[5] A 4-dimensional application of the "Game of Life" can be found in [18].

8.3 Cellular Automata in Algorithmic Composition

Peter Beyls was one of the first to use cellular automata for the generation of musical structure [3]. As an expansion on traditional approaches, Beyls established additional methods for the definition of the neighborhood. Furthermore, a greater number of previous states of a cell may be considered for the calculation of the next state. Musical mapping on tone pitches is carried out by assigning pitches of a scale chosen by the user to the cell positions. Rhythmic values result from the rule that successive identical pitches lead to a held note. The instrumentation of the musical structure is completed for each event by the assignment of a MIDI channel that corresponds to the index of the respective cell modulo the number of the available MIDI channels.

In another work [4], Beyls introduced a 2-dimensional cellular automaton and enables the user to alter different parameters in real time. For the application of rules, in addition to the current state of the neighborhood, previous states may also be considered. The generations in the first CA may be modified by a user through the activation of particular cells[6] or also, by changing the generation rule in regard to the considered radius of the neighborhood. The time-averaged behavior of each cell is saved as a measure of its degree of activity. In order to avoid stationary situations in the CA, cells that are active within a particular time period are deactivated; the duration of this period may be set variably. An additional method that can be used to affect the activity of the CA is to modify the generation rule depending on Chris Langton's Lambda parameter.[7] This value measures the number of cells that transition to an active state in the following generation and indicates a proportional relation between the activated and non-activated cells. Cellular automata with a parameter value between 30% and 40% prove to be particularly productive. The user enters a value for λ – the rule of the CA is then modified until the behavior of the automaton corresponds to the desired activation behavior.

In a later work [5], Beyls described his software CA Explorer which, for the generation of musical structure, uses 1-dimensional cellular automata and Lindenmayer systems that are subsequently further processed by a genetic algorithm.

Beginning in 1990, Dale Millen developed Cellular Automata Music [19, 21], in which the data of 1- to 3-dimensional cellular automata may be mapped in a grid on MIDI values assigning pitch and duration values. In another work [20], Millen described the behavior of an interesting 1-dimensional cellular automaton with two possible states per cell and a neighborhood consisting of five cells ($K2R2$): "The list of rule results may contain instances of repeating sequences of various lengths. Some of these repeating sequences possess an internal structure that can be described as consisting of repetition with variation." [20, p. 398]. This CA is assumed to have a grid size of ten units; each of the 25 distinct states of the neighborhood is indicated by numbers between 0 and 31. In the following example, the cells 4, 5 and

[6] A "cell" in this context – as in some of the following works as well – often also refers to a distinct cell on a particular position in the grid.

[7] Cf. [17], for a detailed description also see [14, p. 242ff].

8 are set to "1" (active) in the initial configuration and the rest of the cells are deactivated. For the musical mapping, only the values of the cells 1, 5 and 9 are considered in further generations. The following rule for the cellular automaton is agreed upon: The neighborhood configurations 2, 6, 7, 11, 12, 15, 16, 17, 21, 22, 26, 27, 30 that represent the neighborhood as five-digit binary numbers cause the activation (1) of the middle cell in the next generation. The table below shows the first generations in an initial configuration of 0 0 0 1 1 0 0 1 0 0. For the musical mapping, the neigh-

	1	2	3	4	5	6	7	8	9	10
Gen0	0	0	0	1	1	0	0	1	0	0
Gen1	0	0	0	1	1	0	0	0	0	1
Gen2	0	1	0	1	1	0	1	0	1	0

borhood configurations of each cell are mapped on MIDI pitches between 60 and 90 (table 8.1). With a neighborhood state of 0, a pause is set. In the table, in generation 1, for example, the resulting value for the mapping of cell 1 is 8 and for the 5th and 9th cell the values are 12 and 2 , according to the neighborhoods 0 1 0 0 0, 0 1 1 0 0 and 0 0 0 1 0. From the first three generations of the observed cells on, regular patterns within the number sequences are produced in the next 40 generations. In order to represent the regularity more clearly, the values of the periodic sequence are divided into 23 entries each, whereas in the first row of table 8.1, the three previously mentioned generations are depicted. In another work [21], Millen described a newer

0	12	8	8	12	2	18	13	20														
1	27	16	17	23	20	6	8	9	5	19	17	24	3	22	26	2	14	4	13	17	24	10
6	27	16	6	23	20	5	8	9	26	19	17	20	3	22	0	2	14	8	13	17	19	10
6	2	16	6	12	20	5	29	9	26	2	17	20	12	22	0	13	14	8	11	17	19	20
6	2	9	6	12	1	5	29	17	26	2	6	20	12	5	0	13	24	8	11	26	19	20
4	2	9	24	12	1	27	29	17	23	2	6	8	12	5	19	13	24	3	11	26	2	20
4	13	9	24	10																		
					1	27	16	17	23	20	6	8	9	5	19	17	24	3	22	26	2	14

Table 8.1 Regularities in the production of a CA.

version of Cellular Automata Music that implements the above mentioned principles in an extended form. The size of the grid is enlarged considerably herein; in addition, it is possible to modify the selection of the columns used for the mapping during playback. For the mapping of the pitches, the user may select from a number of predefined or self-designed scales. Timbre results from assigning MIDI channels to single cell columns. This automaton develops its generations within an adjustable time cycle that determines tempo. The duration of every sonic event depends on the density of the neighborhood. In this version, the software allows for the processing of 1-dimensional cellular automata of the types $K2R1$, $K3R1$, and $K2R2$. The user specifies the CA type and the rule, chooses the size of the grid and selects up to 16 cells of the CA grid that different MIDI channels are assigned to.

An early application of an interactive composition system that is based on cellular automata is developed by Andy Hunt, Kirk Ross and Richard Orton with the software Cellular Automata Workstation. [16]. In this system, 1-dimensional cellular automata with two states and different radiuses may be used for the generation of musical structure. In a simple mapping, the user defines the rule number and selects a particular number of cell positions whose chronological development is transformed into musical output. The automaton starts with the generation of its states at a rate that is set by the user. Ascending from left to right, the cell positions are assigned MIDI values that represent tone pitches – a transposition of the entire tone pitch range may be preset. The pool of pitches for the musical mapping may be chosen variably for distinct segments of generation of the CA, whereas the type of transition ranging from a sudden change to smooth crossfading is also determined by the user. Musical mapping may also be carried out by using values of cell areas as control parameters for synthesis processes.

In 1993, Eduardo Reck Miranda[8] began the development of CAMUS (from Cellular Automata MUSic). This software generates musical structures on the basis of two cellular automata. In this system, the output of Conway's Game of Life and of the Demon Cyclic Space (DCS), a CA created by the mathematician David Griffeath [10], are applied. The term "demon" is to be understood here as a "demon of work," because this CA continuously generates complex patterns that spread reiteratively (Cyclic) on a grid (Space). In a DCS, the cells are initialized with random values between 0 and an indicated maximal value. These values make up the states of the CA and are usually associated with colors. The following rules apply for the behavior of the CA: If a cell has at least one neighbor whose state is higher than its own state by exactly 1, it is adjusted to the neighbor's state. Cells with a maximal value can only be changed by an adjacent cell whose value is 0 and then are assigned a value of 0 as well. Due to the fact that a cell value of maximum +1 changes to 0 again, this rule transfers the idea that cell space behaves exactly like the toroidal grid space.

In Miranda's system, Conway's CA serves to generate pitches and durations and the DCS provides parameters for the instrumentation of the structure. Each cell generated by the automaton receives its musical parameters due to its position in a coordinate system. CAMUS interprets the coordinates as triads, the x-axis representing the interval between a given fundamental tone and the middle chord tone and the y-axis representing the distance between the middle and the upper chord tone (figure 8.10).

After each generation of the cellular automaton, the active cells of the GoL are analyzed sequentially and mapped on the corresponding triads. In order to determine the temporal shape, the program creates a binary list $[s_1, s_2 \ldots, s_8]$ for each triad containing the states of the eight neighboring cells. This list provides the material for four other lists in the following order: w1: $[s_1, s_2, s_3, s_4]$; w2: $[s_4, s_3, s_2, s_1]$;

[8] [22, 23, 24, 25, 26]. Further interesting reading should be found in the newly published book edited by Miranda and Biles [27].

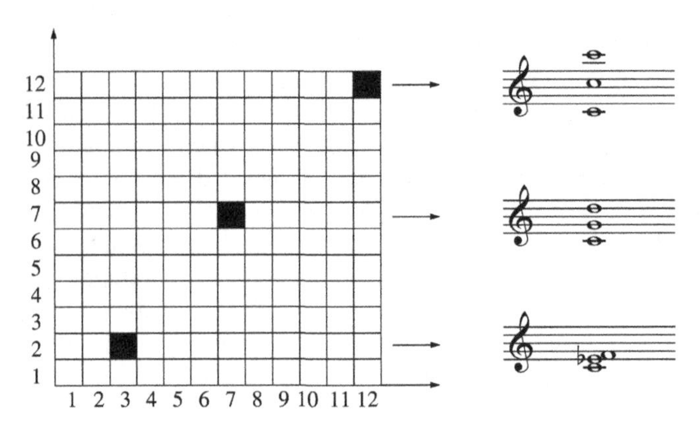

Fig. 8.10 Pitch mapping in CAMUS.

w3: [s_5, s_6, s_7, s_8]; w4: [s_8, s_7, s_6, s_5]. The temporal shape of the triads results from the temporal position of the notes as well as their duration. For the succession and duration parameters, Miranda gives ten configurations that are then combined in pairs (figure 8.11).

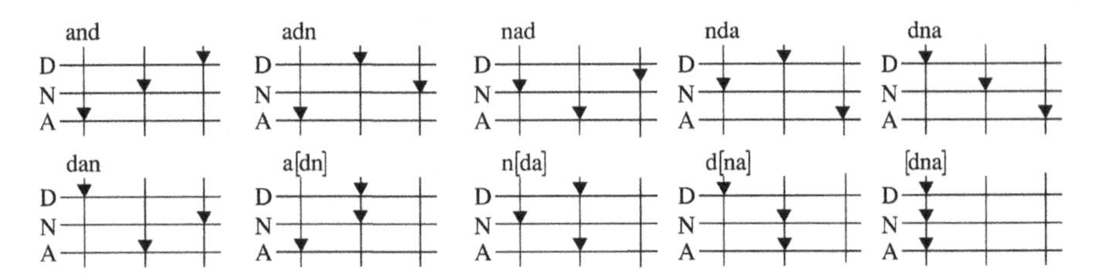

Fig. 8.11 Configurations for the temporal order of the triads.

The temporal order can be obtained by combining two of the abovementioned lists each. For the temporal organization of a triad, a logical operation is used to combine w1 and w2, for the durations, w3 and w4 are combined in the same way. The symbol strings are assigned to the configurations as follows:

0000 a[dn] 0010 adn 0101 and 0111 nad 1011 nda
0001 [dna] 0011 dna 0110 dan 1001 d[na] 1111 n[da]

So, for example with the resulting constellations "nad" and "a[dn]," a time structure is produced for the triad (figure 8.12, left) which may generate concrete musical material in case concrete tone pitches and durations are assigned to the triad structure (figure 8.12, right).[9]

The configuration of the Demon Cyclic Space determines the instrumentation of the composition – the value ranges of the DCS being assigned different MIDI

[9] Miranda does not describe how the dynamic relationships are generated.

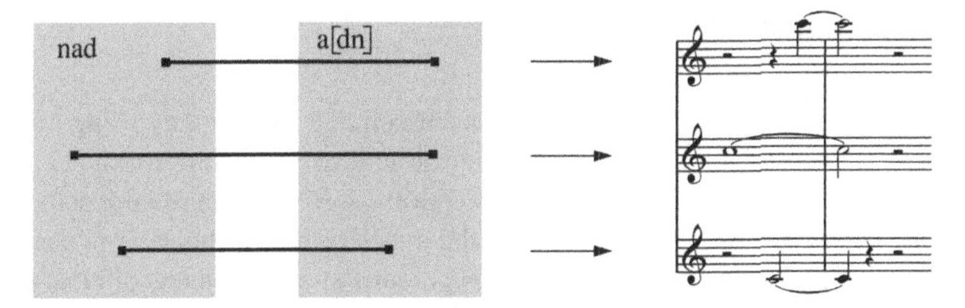

Fig. 8.12 Temporal mapping in CAMUS.

channels; by means of the overlapping of DCS and Gol, the single cells in the GoL are given their own timbres.

In a 3-dimensional version of CAMUS [9, p. 5], the z-coordinate is used for the generation of a further chord part; for the production of the temporal structure, a first-order Markov process is applied.

According to Miranda, CAMUS allows for an efficient application of the behavior of two cellular automata on musical structure: "The results of the cellular automata experiments are very encouraging. They are good evidence that both synthesized sounds[10] and abstract musical forms can be successfully modeled using cellular automata." [22, p. 11]. Even though Miranda's musical conceptions can be successfully realized by CAMUS, the musical mapping, however, deviates from the fundamental behavioral pattern of a CA: In CAMUS, coordinates of active cells are mapped on the pitch structure of triads in a grid of a certain size. Due to this principle, the pitch relationships for each active (meaning sounding) cell are solely determined by the size of the underlying grid. The grid size is, of course, an essential parameter for the graphic representation of a CA, but the basic mode of operation of the CA is determined exclusively by the rule and neighborhood configuration of every single cell. Indeed, the rule of the CA determines which triad constellations are made audible, but the eminently important task of shaping the tone pitches is, however, realized by a peripheral parameter of the CA; this also means that a CA that has an identical initial configuration and the same rule but a different grid size provides wholly different musical results. Due to the fact that "active triads" are additionally read out successively from the grid, a cycle is established that is not caused by the state changes of all cells in the CA as is generally the case, but is only observable in the progression from one cell position to another. While these mapping strategies do not conform to the basic behavior of a CA, this is not necessarily a weakness; it may be considered a legitimate variant to make one's own musical preferences audible by means of a system of algorithmic composition. In this context it is also interesting that the instrumentation of the triads is produced by the overlapping of two distinct types of CA, thus intentionally forgoing a structural relation between harmonic and temporal organizations of the triads on the one hand, and their timbral aspect on the other hand.

[10] Miranda refers here to another of his software programs, Chaossynth; cf. [23, p. 3–4].

8.3.1 Polyrhythmic Structures

Alan Dorin has developed interesting applications of cellular automata by using his system Boolean Sequencer for producing rhythmic structures. In one approach [11], Boolean networks (BN) are used as structure generators. A BN is a set of connected nodes; the state of each node is determined by the states of its neighboring nodes, which are designated as the node's input. The BN is represented by means of binary symbol strings; all nodes change their states simultaneously. Dorin applies bars that correspond to a generation in a 1-dimensional CA. In contrast to a traditional CA, for every cell position which is represented here by its position in the bar, a particular rule in the form of a logical function may be selected. Therefore, these logical functions determine the state of each note on each bar position by the neighborhood of the notes on the corresponding positions in the previous bar. The OR function activates the note if either one of the two or both neighbors of the previous note were active. The XOR function activates the note if one but not both neighbors of the previous note was active. In this system, NOT activates the note if neither of the two neighbors was active in the previous state. The AND node requires both of its neighbors to be active in the previous state of the BN. The activated notes of a bar are played successively on their temporal position, the BN moves depending on the defined rules to its new state and the process begins again. In Dorin's model, the Boolean networks are implemented as a multitrack system whose tracks are represented by the single bars. Different rules as well as a particular MIDI channel may be assigned to each track. During playback, pitch, duration and velocity of the nodes may be altered.

In another software implementation LIQUIPRISM [12], Dorin used a cube whose six sides each correspond to the grid of a cellular automaton. For each cellular automaton it may be determined how fast it produces new generations. The cells of each CA which border an edge of the cube, also count the closest cells on the adjacent faces as their neighbors. Hence, the usual toroidal structure of the CA is dissolved and transferred to the topological form of a sphere. For the musical mapping, each cell is assigned to a particular tone pitch and the sides of the cubes to different MIDI channels. A note generates a sonic event when the state of the respective cell changes from 0 to 1. A threshold value may be set that limits the number of notes that are made audible simultaneously.

8.3.2 Comparison of Cellular Automata

Eleonora Bilotta and Pietro Pantano [7] examined the behavior of different 1-dimensional cellular automata, applying a musical mapping for the purpose of a sonification of a dynamic process rather than to the generation of interesting musical structure. With the so-called "local musification codes," the cells of the CA are transferred to pitches: If K (the states) is 2, one tone pitch is determined for each cell, whereas in $K > 2$, each state value of the cell is given another tone pitch. In producing the tone pitches in this way, this code reads the CA line by line from left to right. By doing that, the usually common simultaneous state change of all cells of the CA in musical mapping is not considered. "Global musification codes" calculate the total activity in every generation and make it audible by means of musical mapping. "Mixed musification codes," in contrast, use segments of a 2-dimensional representation of numerous generations of a 1-dimensional cellular automaton and assign musical parameters to these parts. In another work by Bilotta and Pantano[8], genetic algorithms are applied for the selection of optimal CA rules. As a fitness function, in the melodic output of the CA, the consonance is measured in intervallic progressions. Furthermore, repeating structures in terms of motifs are searched for as well.

In cooperation with Valerio Talarico, Eleonora Bilotta and Pietro Pantano [6] examined 1-dimensional automata within Wolfram's four classes in regard to their behavior under different rules and the production of periodical appearances of graphic patterns. According to the authors, the sonified results show that the different observed CA structures are highly distinguishable; concrete examples of the sonifications, however, are not given.

8.4 Synopsis

Cellular automata are often mentioned together with genetic algorithms within one category. While the terminologies for both of these methods suggest a close relation with biological processes, they do not share any common functional principles. Cellular automata may, however, be compared to substitution systems and above all context-sensitive Lindenmayer systems that produce symbol strings of the same length in each generation and by doing this may be equivalent to a 2-dimensional cellular automaton [13]. Apart from these similarities, the basic differences are, however, evident. Lindenmayer systems are producing an increasing number of symbols. Because of their self-similarity the output is easily predictable. A frequent mapping represents active cells as visible; therefore, if a large number of cells also change their state to "active" during the CA generations, it suggests an increasing number of cells. But naturally, the number of cells in a CA remains constant. Moreover, the aspect of self-similarity is – in the same way as it is fundamental in a Lindenmayer system – possible in some configurations of cellular automata,[11] but is rather an exception than the rule. Properties essential for the operation mode of a traditional CA are infiniteness and also a toroidal structure of the underlying grid as well as the irrelevance of the local position of each cell, in case the configuration of the immediate neighborhood is not considered. Despite the simple rules and the deterministic character, the behavior of most of the cellular automata is extremely complex and difficult to predict in the resulting configurations.

[11] Cf. the Z1R1 automaton of class four and rule 90, see 1-dimensional automata.

A musical mapping that may be observed frequently does not follow the basic principles of the CA, but instead uses a mapping of the graphic representation, which itself already constitutes a mapping of the actual algorithm. In the graphic representation, the cells seem to move over the grid – an impression that, however, is only produced by the state changes of all cells of the CA. If values of musical parameters are assigned to particular positions in the grid, and consequently, a particular cell state on these positions leads to the output of the respective values, the principle of irrelevance of the local position of the cells is not taken into account in the musical mapping since the configuration of the neighborhood is the only decisive factor for the state changes. This kind of musical mapping therefore interprets the grid of a CA as a musical event space that is crossed by moving cells. Here, a monotonously ascending or descending output is reduced by means of a modulo operation to the range of a particular parameter – a procedure which is necessary in case of a limited number of tone pitches. Due to this mapping, another dissonance in terms of the toroidal structure of the CA grid may also occur. Because in most cases a scale is given, for example, for the tone pitches on the grid positions along a CA axis – naturally, a scale consisting of dynamic values or other musical parameters could also be used – the border areas of the grid represent the lowest or highest tones. If in the abovementioned mapping, the cells cross the visual border of the grid, thus wrapping around to the other side, a maximal change of tone pitch results, even though the toroidal structure of a CA would actually only cause a change by one scale step.

These works, however, also raise the question as to whether the characteristic behavior of a particular algorithmic class must be reflected by the way of musical mapping. Of course, in each system of algorithmic composition, the musical output is of paramount importance; if, however, the specific properties of the applied algorithms are not suitably used for structure generation, the motivation for selecting them is questionable – these applied algorithms become exchangeable and one runs the risk of generating arbitrary material. The fact that the characteristics of an algorithm may indeed be creatively modified in order to achieve specific musical results is shown amongst others by the work of Dorin, who transforms several 2-dimensional grids to the topological form of one sphere. The velocities of the CA cycles are furthermore assigned to the grid areas and interpreted as metric layers; this strategy of musical mapping allows a basic principle of the CA – the cyclic actualization of all cell values – to become audible in a polyrhythmic structure.

Just like Lindenmayer systems, cellular automata are an exception among the different paradigms of algorithmic composition because they are almost exclusively used for the implementation of personal compositional strategies. The generation of style imitations is rather unfeasible as it is impossible to analyze and encode a corpus within a CA; nevertheless it is, however, theoretically conceivable to generate musical material by means of a CA's rules which may, due to its homogeneity, correspond to a particular stylistic notion.

References

1. Bays C (1987) Candidates for the game of Life in three dimensions. Complex Systems 1, 1987
2. Bays C (2005) About trilife cellular automata. University of South Carolina Department of Computer Science and Engineering
   http://www.cse.sc.edu/ bays/trilife3/Readme.html Cited 18 Jul 2005
3. Beyls P (1989) The musical universe of cellular automata. In: Proceedings of the 1989 International Computer Music Conference. International Computer Music Association, San Francisco
4. Beyls P (1991) Self-organising control structures using multiple cellular automata. In: Proceedings of the 1991 International Computer Music Conference. International Computer Music Association, San Francisco
5. Beyls P (2003) Selectionist musical automata: Integrating explicit instruction and evolutionary algorithms. In: Proceedings of IX Brazilian Symposium on Computer Music, 2003
6. Bilotta E, Pantano P, Talarico V (2000) Music generation through cellular automata: How to give life to strange creatures. In: Proceedings of Generative Art Conference GA2000, Milano, Italy
7. Bilotta E, Pantano P (2001) Artificial life music tells of complexity. In: ALMMA 2001: Proceedings of the workshop on artificial life models for musical applications. Linguistics Department, University of Calabria, Arcavacata di Rende
8. Bilotta E, Pantano P (2002) Synthetic harmonies: An approach to musical semiosis by means of cellular automata. Leonardo, 35/1, 2002
9. Burraston D, Edmonds E (2005) Cellular automata in generative electronic music and sonic art: Historical and technical review. Digital Creativity 16/3, 2005
10. Dewdney AK (1989) Cellular universe of debris, droplets, defects and demons. Scientific American, August, 1989
11. Dorin A (2000) Boolean Networks for the generation of rhythmic structure. In: Proceedings of the 2000 Australian Computer Music Conference
12. Dorin A (2002) LIQUIPRISM: Generating polyrhythms with cellular automata. In: Proceedings of the 2002 International Conference on Auditory Display, Kyoto, Japan, July 2–5, 2002
13. DuBois RL (2003) Applications of generative string-substitution systems in computer music. Dissertation. Columbia University, 2003
14. Flake GW (1998) The computational beauty of nature. Computer explorations of fractals, chaos, complex systems, and adaption. MIT Press, Cambridge, Mass
15. Gardner M (1970) Mathematical games: The fantastic combinations of John Conway's new solitaire game "live." Scientific American, 223, October, 1970
16. Hunt A, Kirk R, Orton R (1991) Musical applications of a cellular automata workstation. In: Proceedings of the 1991 International Computer Music Conference. International Computer Music Association, San Francisco
17. Langton C (1986) Studying artificial life with cellular automata. Physica D 22, pp 120–149
18. Meeker Lee (1998) Four-dimensional cellular automata and the Game of Life. Thesis, University of South Carolina
19. Millen D (1990) Cellular Automata Music. In: Proceedings of the 1990 International Computer Music Conference. International Computer Music Association, San Francisco
20. Millen D (1992) Generations of formal patterns for music composition by means of cellular automata. In: Proceedings of the 1992 International Computer Music Conference. International Computer Music Association, San Francisco
21. Millen D (2005) An interactive cellular automata music application in cocoa. In: Proceedings of the 2004 International Computer Music Conference. International Computer Music Association, San Francisco
22. Miranda ER (2001) Evolving cellular automata music: From sound synthesis to composition. In: ALMMA 2001: Proceedings of the workshop on artificial life models for musical applications. Linguistics Department, University of Calabria, Arcavacata die Rende

23. Miranda ER (2003) On the music of emergent behaviour: What can evolutionary computation bring to the musician? Leonardo, 36/1, 2003
24. Miranda ER (2003) On making music with artificial life models. 5th Consciousness Reframed Conference, University of Wales College, Newport, Caerleon, Wales, UK, 2003
25. Miranda ER (2003) Introduction to cellular automata music research. DIGITAL music online tutorials on computer music.
http://x2.i-dat.org/ csem/UNESCO/8/8.pdf Cited 29 Mar 2005
26. Miranda ER (2003) At the crossroads of evolutionary computation and music: Self-programming synthesizers, swarm orchestras and the origins of melody. Evolutionary Computation, 12/2, 2003
27. Miranda ER, Biles JA (eds) (2007) Evolutionary computer music. Springer, London. ISBN 13 978-1-84628-599-8
28. Von Neumann J (1966) Theory of self-reproducing automata. University of Illinois, Urbana, Ill
29. Rennard JP (2006) Introduction to cellular automata.
http://www.rennard.org/alife/english/acintrogb01.html Cited 2 Aug 2007
30. Siver S (2003) Life Lexicon.
http://www.argentum.freeserve.co.uk/lex_z.htm Cited 14 Dec 2004
31. Toffoli T, Margolus N (1987) Cellular automata machines: A new environement for modeling. MIT Press, Cambridge, Mass. ISBN 0262200600
32. Wojtowicz M (2004) Mcell 4.20.
http://www.mirekw.com/ca/index.html Cited 13 Dec 2004
33. Wolfram S (2002) A new kind of science. Wolfram Media, Champaign, Ill. ISBN 1-57955-008-8
34. Zuse K (1970) Calculating space. MIT Technical Translation, MIT Press, Cambridge, Mass

Chapter 9
Artificial Neural Networks

Artificial neural networks (ANN) enable problem solving by changing the structure of interconnected components. In analogy to the biological model, these interconnected elements are referred to as *neurons* and make up the basic units of information processing. A simple model of a biological neuron is composed of the following components, illustrated in figure 9.1: A large number of short nerve fibers called *dendrites* conduct impulses from the adjacent neurons to the cell body. The cell reacts to this input with an impulse that is carried to other cells by an *axon*. An axon is a long nerve fiber that may be up to one meter in length and undergoes extensive branching at its end which enables communication with dendrites or other cell bodies. The neurons are connected via *synapses*. When an electrical impulse runs through an axon, this causes the synapse to release a substance called a neurotransmitter which is picked up by a neighboring dendrite and changes the electric potential of the cell. When a particular threshold value is reached, an *action potential* is generated in a cell that in turn is sent along the axon through the synapses to the dendrites of the connected neurons. The human brain has approximately 10^{11} interconnected neurons whose reaction time is estimated at one millisecond. The average processing time for the recognition of complex visual and/or acoustic information by a human brain is approximately 0.1 seconds. If biological neuronal systems would work like strictly sequential digital computers, only 100 instructions would be possible in that time frame [12, p. 172]. To overcome this restriction, knowledge processing is performed in both the biological model and the artificial neural network in parallel, meaning that all neurons of one level process their outputs at the same time.

Artificial neural networks are used in numerous applications; depending on the type of ANN, they are often applied in pattern recognition, prediction, optimization and automatic classification.

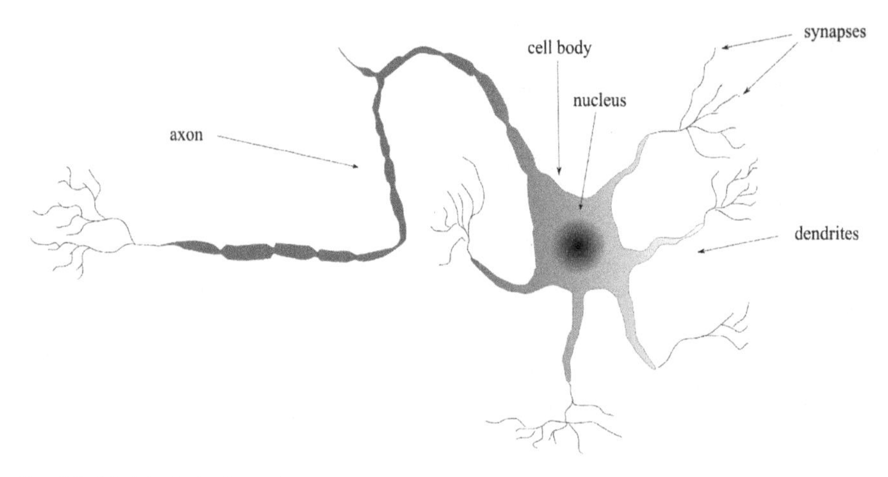

Fig. 9.1 Model of a biological neuron.

9.1 Theoretical Basis

The fundamental element of every ANN is the artificial neuron that in basic models is composed of the following components: An interface for taking up information from the environment or the output of other connected neurons; a *propagation function* that combines all inputs of the neuron into one piece of information; an *activation function* which, considering a threshold value, determines the activation state of the neuron on the basis of the value of the propagation function. The *output function* calculates an output from the activation state of the neuron that is sent to all connected neurons. The state of activation and the threshold value of the neuron are stored in a local memory. The activation functions used are, depending on the type of learning method, different *sigmoid* (resembling an "S") functions. Simple variants are the linear function or the threshold function; in a linear function (figure 9.2, top left), the input information is adopted directly, whereas the threshold function (figure 9.2, top right) only allows for an activation of the neuron when the present net input exceeds the threshold value. If the steepness of the curve is to be influenced by an additional parameter (c), the *hyperbolic function* (figure 9.2, bottom left) or the *logistic function* (figure 9.2, bottom right) may be applied as well.

The arrangement of the neurons in the respective model determines the architecture or topology of the ANN whose different forms are described later in detail. As information is processed within the network structure in numerical values, appropriate encoding schemes must be designed in order to enable the ANN to interpret input and output values. An encoding scheme is also decisively responsible for the implementation of an appropriate learning algorithm.

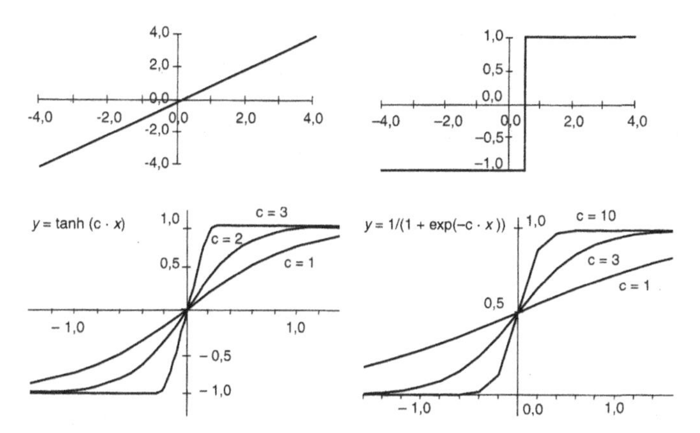

Fig. 9.2 Activation functions of an artificial neuron [12, p. 175]. With kind permission of Carl Hanser Verlag.

9.2 Historical Development of Neural Networks

The first reflections on connectionist structures were made in 1943 by the neurophysiologist Warren St. McCulloch and the mathematician Walter Pitts, who dealt with the calculation of reaction patterns in nervous systems.[1] In another work, McCulloch describes a non-hierarchic structure in nervous systems that he discovered by means of studies on the spinal marrow of frogs.[2] The so-called *McCulloch–Pitts neuron*, as the earliest example of a connectionist model, enables the calculation of logic functions. In 1949, the psychologist Donald Olding Hebb formulated a principle which would be one of the first learning rules for connectionist architectures. According to this principle, the weights of frequently used connections are increased. From a biological point of view, Hebb states: "When an axon of cell A is near enough to excite cell B and repeatedly or persistently takes part in firing it, some growth process or metabolic change takes place in one or both cells such that A's efficiency, as one of the cells firing B, is increased."[3] This principle is referred to as *Hebbian learning* and is implemented in artificial neural networks of different architectures as well as used both in supervised and unsupervised learning. In 1957, Frank Rosenblatt began with the development of another model of ANN, the *perceptron*. At the same time, he was supervising the construction of the first fully functioning neurocomputer[4] in the Cornell Aeronautical Laboratory, the Mark I Perceptron. In 1962, Rosenblatt described theoretical elements and possible applications

[1] "A Logical Calculus of the Ideas Immanent in Nervous Activity" [14].

[2] "A heterarchy of values determined by the topology of nervous nets" [15].

[3] From "The Organization of Behavior" [3], cited after [13, p. 690].

[4] Already in 1951, Marvin Minsky, the founder of the Artificial Intelligence Laboratory at the MIT, developed the neurocomputer "Snark" in the course of his doctoral thesis in Princeton. However, the model was never built.

of the perceptron[5] and showed that this model was also able to learn every function it could represent by entering weights and threshold values.[6] In 1959, Marvin Minsky and Seymor Papert pointed out the restrictions and limitations of this model[7] which led to a more critical view of connectionist problem solving in the following years and also to the fact that symbol-based methods were awarded a universal claim to problem solving. In 1960 Bernard Widrow and Marcian E. Hoff[8] developed the *Adaline* (short for: *Adaptive linear element*), an extended model of the perceptron which generates outputs of +1 and –1 and, similar to the McCulloch–Pitts neuron, has an additional input. Teuvo Kohonen's *linear associator* from 1972 represents an extension of the Adaline as it has several neurons in the output layer. This type of architecture is also referred to as *Madaline* (*Multiple Adaline*).[9] In 1974, in his doctoral thesis[10] at Harvard University, Paul Werbos laid the foundation for the *back-propagation algorithm* that became popular through publications by David Rumelhart, Geoffrey Hinton and James McClelland [21, 22] in 1986. Beginning in 1982, John Hopfield created a connectionist architecture named after himself that are described in two works [6, 7]; in the same year, Kohonen developed the principle of *Kohonen feature maps* (also: self-organizing maps, SOM).[11] From 1976 on, Stephen Grossberg and Gail Carpenter created the *adaptive resonance theory* (*ART* or also *ARTMAP*), a neural network architecture consisting of connectionist models for automatic classification.

9.3 The Architecture of Neural Networks

The McCulloch–Pitts neuron (figure 9.3) is the first model of an artificial neuron whose different switching possibilities generate the basic architecture of an ANN. The information processing in this model is explained in the following example by means of two logic functions [13, p. 665]. The inputs are excitatory (+1) or inhibitory (–1), a third input is called *bias* or *calibration* and in this case has a constant value of 1. When the AND function is calculated, the neuron gets the weightings +1, +1, –2 for the three inputs, multiplies each weight by the respective input and sums up the results. If a value is greater than 0, the condition is fulfilled and the neuron outputs the value 1. When the weighting for the calibration is set at –1, the neuron yields the correct results for the OR function.

Figure 9.4 illustrates different types of networks by means of matrices and graphs. When a neuron is neither part of the input layer nor the output layer, and

[5] "Principles of Neurodynamics" [20].

[6] Also known as "perceptron-convergence theorem."

[7] "Perceptrons: An Introduction to Computational Geometry" [16].

[8] "Adaptive switching circuits" [30].

[9] "Correlation Matrix Memories" [9].

[10] "Beyond Regression: New Tools for Prediction and Analysis in the Behavioral Sciences" [29].

[11] "Self-organized formation of topologically correct feature maps" [10].

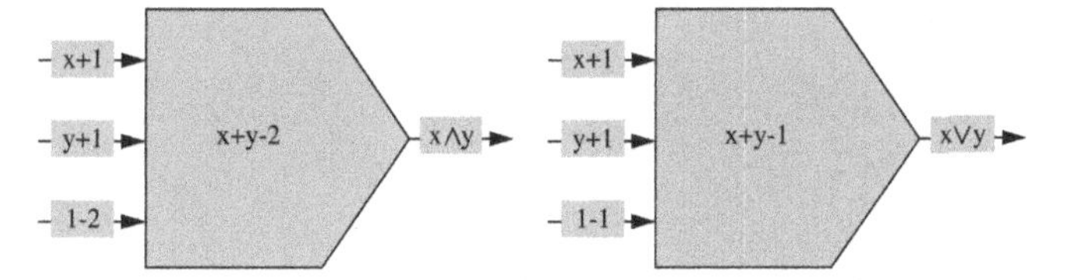

Fig. 9.3 Logic calculation in the McCulloch–Pitts neuron.

is consequently only connected to other neurons, it belongs to the so-called *hidden layer*. A *feedforward network* (figure 9.4, top left), of which the perceptron is an example, is characterized by an input and an output layer, as well as one or more hidden layers. The denomination feedforward is explained through the fact that the connections only point in one direction, from inputs through hidden layers to outputs. A model of this network type with numerous hidden layers (figure 9.4, top right) also requires a more complex training algorithm. This model is also known as *multi-level perceptron*. The other structures illustrate network types that skip one or more layers (figure 9.4, middle left), recurrent structures (figure 9.4, middle right), and completely interconnected structures that are not directly recurrent (figure 9.4, bottom). Within this classification, those network architectures must also be mentioned in which processing is performed within one single layer, such as is the case in *self-associative* or *Hopfield nets*. The adaptive resonance theory, as an adaptive network architecture, additionally involves information on previous network patterns in the problem solving process as well. Further possibilities for classifying ANNs are given by dividing up the networks into trainable and non-trainable nets, as well as making a distinction in terms of the type of learning algorithm.

9.3.1 The Perceptron

The perceptron (figure 9.5) is a model of a feedforward network that is constructed in regard to visual pattern recognition based on the concept of an artificial eye. The perceptron of a simple form has no inner layer and the information from the image is passed to the input layer by non-trainable connections and fixed weights and from the input layer to the output layer with trainable connections and adjustable weights. In the binary model of the perceptron, inputs and outputs may only assume binary values and the weights are represented in real numbers. A simple threshold function is applied here and the output layer may consist of one or more neurons. As a training algorithm, the perceptron uses the delta rule which is a form of supervised learning in which the weightings of the neurons are updated corresponding to the error in the output of the perceptron [13, p. 672ff]. An extended model of a perceptron has a number of trainable layers and is known as a *multi-level perceptron* or also – corresponding to the learning algorithm that is applied in this type – a *back-propagation net*.

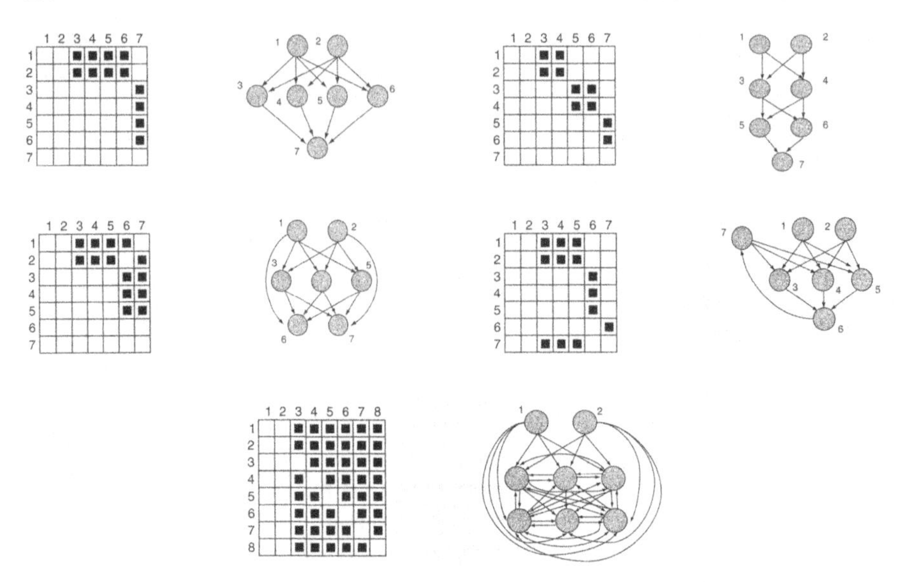

Fig. 9.4 Different types of artificial neural networks [12, p. 178–179]. With kind permission of Carl Hanser Verlag.

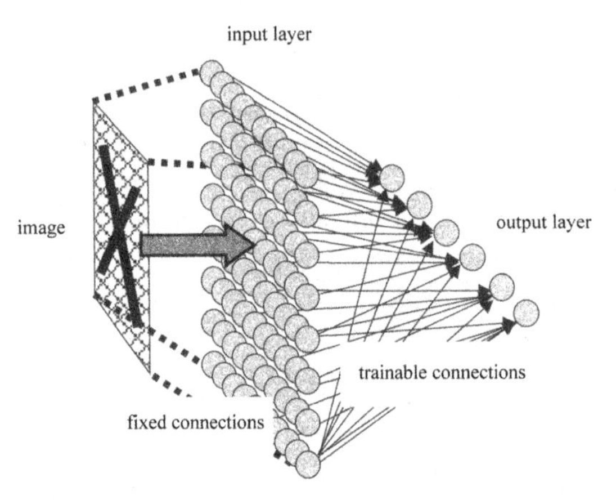

Fig. 9.5 Perceptron for pattern recognition [12, p. 184]. Terms in figure translated into English. With kind permission of Carl Hanser Verlag.

The abovementioned criticism by Minsky and Papert regarding the possible applications of a perceptron referred to the one-layer model that is unable to find solutions for non-linear separable functions, and this diminished interest in ANN for some time. This means that if the outputs of the net are represented within a two-dimensional coordinate system, correct and incorrect results cannot be divided into two half-planes by a straight line.

9.3.2 The Back-Propagation Net as an Extension of the Perceptron

A back-propagation net is a perceptron extended by additional layers and solves the class of non-linear separable functions. This type of ANN is trained with the back-propagation learning algorithm that also represents a method of supervised learning. The back-propagation algorithm[12] changes the connection weights depending on the net error, which is the difference between the expected and the actual output of the ANN. The weights are changed beginning with the connections to the output layer and then continue backwards to the connections of the input layer. Two-level binary perceptrons are able to produce polygons for classification through the superposition of planes by subtending the planes with an additional AND neuron. In a three-level perceptron, any number of sets for classification may be generated by superposition of polygons; additional levels within the perceptron do not allow further differentiation.

9.3.3 Recurrent Neural Networks

If information from a previous pass is important for the actual one, *recurrent neural networks* can be used, which implement feedback in their design, like by connections leading from units in a layer to the same layer or to previous ones. In a *Jordan net*, this happens by means of *context neurons* that transfer the information of the output layer again to the inner layer (figure 9.6, left); here, the number of context neurons corresponds to the number of neurons in the output layer. This ANN is trained with the back-propagation algorithm. In an *Elman net* (figure 9.6, right), recurrent processing is done by context neurons within the inner layer; the number of context neurons equals the number of neurons in the inner layer.

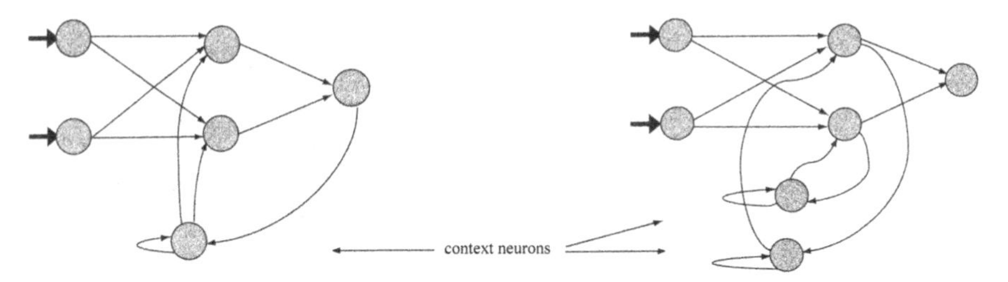

Fig. 9.6 Scheme of a Jordan (left) and an Elman (right) net [12, p. 224, 226]. With kind permission of Carl Hanser Verlag.

---

[12] [13, p. 675ff], [12, p. 190ff].

9.3.4 Kohonen Feature Maps

For classification tasks in the context of unsupervised learning, Kohonen feature maps are used. Within this architecture, two structural levels are applied: An *input layer* and a *map layer*, each neuron of the input layer being connected to each neuron of the map layer. The neurons within the map layer are in turn completely linked to each other, but not to themselves. For automatic classification, a series of data is repeatedly presented to the net in unordered form. The neurons of the map layer are initialized with random weights; in a further step, a so-called *winner neuron* is established whose state is active for all those input patterns that resemble each other in one way or another. During the learning process, the weights of the winner neuron and the circumjacent neurons are adapted to the input patterns. After a number of cycles, local maxima that represent positions for similar datasets will form in the map layer.

9.3.5 Hopfield Nets

In contrast to Kohonen feature maps, which may be applied to recognize structures in unordered inputs, self-associative or Hopfield nets are used for the allocation of input material in terms of given categories. A Hopfield net consists of one single neuron layer within which the neurons are linked completely to one another. The initial weighting of the net is given by a pattern which must be recognized; a cyclic process decides after several steps if the input pattern matches the pattern to be recognized within certain tolerance ranges.

9.3.6 The Adaptive Resonance Theory

Automatic classification may also be performed by different network architectures of the adaptive resonance theory. Patterns are assigned by examining the similarities of these two previously categorized classes. If a large deviation to the already classified areas is detected in a new pattern, a new category is generated. The ARTMAP was originally developed to solve the *stability-plasticity dilemma*. In this context, *stability* means that once recognized patterns are not covered or deleted by new patterns; this would mean that in a new, significantly differing pattern the characteristics of the class which the pattern is assigned to is changed extensively. *Plasticity* refers to the ability to generate new classes in order to maintain stability. The stability-plasticity dilemma results from the fact that it is not always that easy to decide if the new pattern, based on its deviation, only belongs to an already classified class or if a new class should be generated.

9.4 Artificial Neural Networks in Algorithmic Composition

Hermann Hild, Johannes Feulner and Wolfram Menzel [4] developed HARMONET, a system that harmonizes melodies in the style of J.S. Bach based on neural networks and a rule-based system. For the generation of a harmonic skeleton, a recurrent net with a hidden layer is used. This architecture is consequently extended to three parallel nets. An interesting representation form in HARMONET indicates each tone pitch as the set of all harmonic functions that may contain this pitch as a chord element (table 9.1). In the generation of the harmonic skeleton, each quarter note is

Fct.	T	D	S	Tp	Sp	Dp	DD	DP	TP	d	Vtp	SS
C	1	0	1	1	0	0	0	0	0	0	0	0
D	0	1	0	0	1	0	1	0	0	1	1	1
E	..											

Table 9.1 Representation of tone pitches in HARMONET.

harmonized, taking into account its local context that functions as the input of the net (figure 9.7).

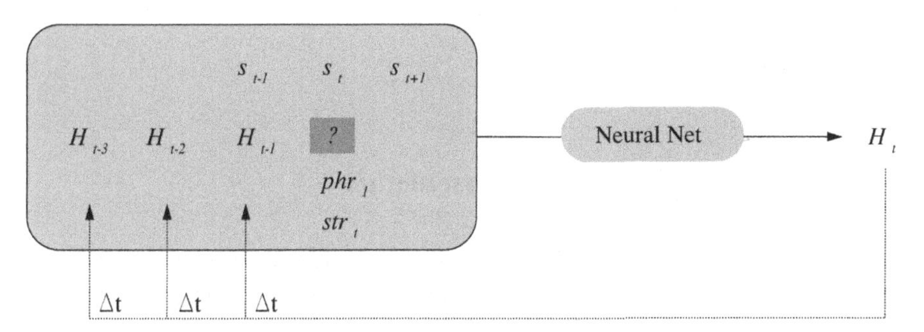

Fig. 9.7 Local context for harmonic skeleton in HARMONET.

The current harmony (H_t) consists of the harmonic (H_{t-1},H_{t-2},H_{t-3}) and the melodic (s_{t-1},s_t,s_{t+1}) context; phr_t contains information on the position of H_t relative to the beginning or the end of a musical phrase; str_t as a Boolean value indicates whether the current harmony is a stressed quarter. For the encoding of this context, 106 neurons are used. 70 of these neurons are used within a hidden layer and 20 form the output of the net. Having been generated by the ANN, the harmonic progressions are controlled by constraints that examine the material in terms of voice distances, parallels and the like. The net is trained on two sets of Bach chorales, each containing 20 chorales in major and minor respectively, using the back-propagation algorithm as training method. In a final step, ornamenting eights are added to the chord skeleton by a further net structure which also works in a context-based manner. In an extended version of HARMONET, three parallel nets are used to generate the harmonic skeleton: These nets work with variable context sizes and the harmonization produced most frequently is transferred to other nets that determine chord inversions and dissonances of the harmonization.

Fig. 9.8 Example of a harmonization by HARMONET.

9.4.1 Strategies of Musical Representation

In an early work by Peter M. Todd [28], basic considerations on the application of neural networks in algorithmic composition were made.[13] An essential question therein is the factor of time and how it can be represented. In one case, a window may be laid over a number of musical events (e.g. notes in a measure). The network receives a measure as input and produces a measure as output. In this "spatial" representation, the different time points (successive notes) are represented by single neurons – time is virtually mapped on space. In the second alternative that corresponds more to the treatment of time in musical processes, a net is used that produces successive notes; the output of a note depends on a number of previously produced notes. A network of this type requires a memory for past events, meaning that it is possible to apply notes of the output as input for the generation of the next note. Therefore, recurrent networks must be used that are able to store a particular number of already produced events. The behavior of the two networks differs: The "spatial" network generates a particular output on the basis of an input, whereas the

---

[13] Further interesting reading by Todd should be found in the newly published book edited by Miranda and Biles [17].

"sequential" network continuously produces new events due to its recurrent connections.

For the production of melodies, Todd uses a recurrent Jordan net with three layers (figure 9.9). The output of the net consists of notes with associated information about pitch and duration. The *plan units* are neurons that designate the melody that is currently processed; *context units* maintain notes produced so far. The *output units* generate the current notes, every output neuron being assigned a context neuron. The neurons of the hidden layer that are connected with the plan units, context units as well as with the output layer by a set of learned, weighted connections, represent the compact encoding for the *generalization capability* of the net, meaning the ability of a neural network to generalize the knowledge represented by the training set – in this case this refers to the possibility to generate similarly designed musical material. The current output is finally transferred to the beginning of the recurrent cycle.

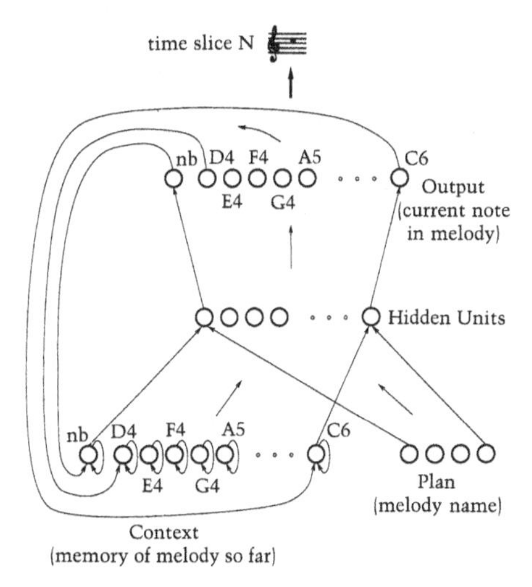

Fig. 9.9 Jordan net for melody generation [28, p. 30]. © 1981 by the Massachusetts Institute of Technology.

For the representation of tone pitch and duration, different approaches may be used. In regards to the pitch, a "relative" (intervallic) or an "absolute" (designating concrete pitches) representation may be chosen. The "relative" representation, in principle, may lead to a serious disadvantage. As soon as the net error produces one single "wrong" tone pitch, this affects all succeeding pitches due to transposition. Therefore, in the context of this ANN, an "absolute" representation is preferable. Within this form, representation may occur again in two different ways. In "distributed" representation, numerous neurons of the output layer represent one tone pitch; within a "localist" representation each pitch is assigned a particular output neuron (an example for "localist" representation with three neurons would be: 0 0 1 = C, 0 1 0 = C#, 1 0 0 = D – the "1" indicating the activation – therefore: 0 0 0 = rest). A "distributed" representation may be produced by a binary encoding of the activation states (e.g. 1 0 0 = C, 1 0 1 = C#, 1 1 1 = D). However, the "distributed" representation proves to be disadvantageous: Due to the fact that for the representation of notes (that are represented as equal), different numbers of neurons are used, it is difficult to train the ANN in a "distributed" representation. So, the net error is, by choosing a note C (one neuron) instead of D (three neurons), higher than, for example, with C# (two neurons) instead of D (three neurons), although in both of these cases, only one "wrong" note is produced; this means that equally "serious" errors are evaluated differently. As for the pitch representation, the duration of notes may also be presented in two different ways. First, a separate pool of output and context units, alongside with the output units for the pitch information, can be used, where this representation can also be "local" or "distributed." In the other alternative, which Todd ultimately chose for his concept, the melody is divided up into slices of time of equal length. Each output in the sequence corresponds to a pitch during one slice of time, which also represents the smallest rhythmical unit. Numerous notes of the same pitch may either produce longer rhythmic values by being tied or tone repetitions. The information "nb" (note-begin) (figure 9.9, bottom left) indicates the beginning of a new note for the necessary differentiation of both cases. After a training process with melodic material, the ANN generates melodies as shown in figure 9.10.

Fig. 9.10 Melodies generated by a recurrent ANN.

The productions of the ANN are often limited in the output to stationary repeating melodic phrases (repeat mark in figure 9.10). Todd also compares the efficiency of his ANN to productions of Markov processes. A great difference from a Markov process lies in the treatment of the context of a note value, be that related to the tone pitch or duration. In a MM, the probability of the production of a certain note value depends on transition probabilities of note sequences in a corpus. The context depth given by the order of the Markov process, however, follows a sequential consideration of past events. So, the statistical prediction of a note value in a Markov process of *n*th order is only possible if the already produced n note values also occur in the same order in the underlying corpus. An ANN, however, is not bound to this strictly sequential view. Even though the consideration of a particular – although generally not very large – context is possible, note values may still be generated that do not occur in the same sequential order in the corpus. The advantage of this freer context treatment of an ANN is also explicitly mentioned in the following work which, in a similar comparison, contrasts Markov processes as well as the formalism of generative grammar with an ANN. Nevertheless, this special type of context treatment, naturally, does not have to be an explicit advantage because choosing a particular method will mainly motivated by the composition of the training corpus as well as the desired properties of the structure to be produced.

Michael C. Mozer [18] developed an ANN that, among other things, enables the production of melodies with underlying harmonic progressions. The system CONCERT is trained on soprano voices of Bach chorales, folk music melodies and harmonic progressions of various waltzes. The architecture of the software is based on a recurrent Elman net and for the training, a modified back-propagation algorithm is used.[14] This system differs from the previously mentioned approaches in terms of the interpretation of the ANN's output. In most of the applications of neural networks in algorithmic composition, the outputs are interpreted as concrete note values. In CONCERT, the output of the net can consist of probabilities for the selection of the next note values, whereas a variable setting of the weights within the recurrent structure allows for their adaptation to the properties of the training set.

The specialty of this ANN lies in its representation of tone pitch, duration and harmonic function through a multi-layer representation model. Conventional methods present a certain disadvantage especially in regard to the perceptual similarities of certain pitches due to psychological reasons.

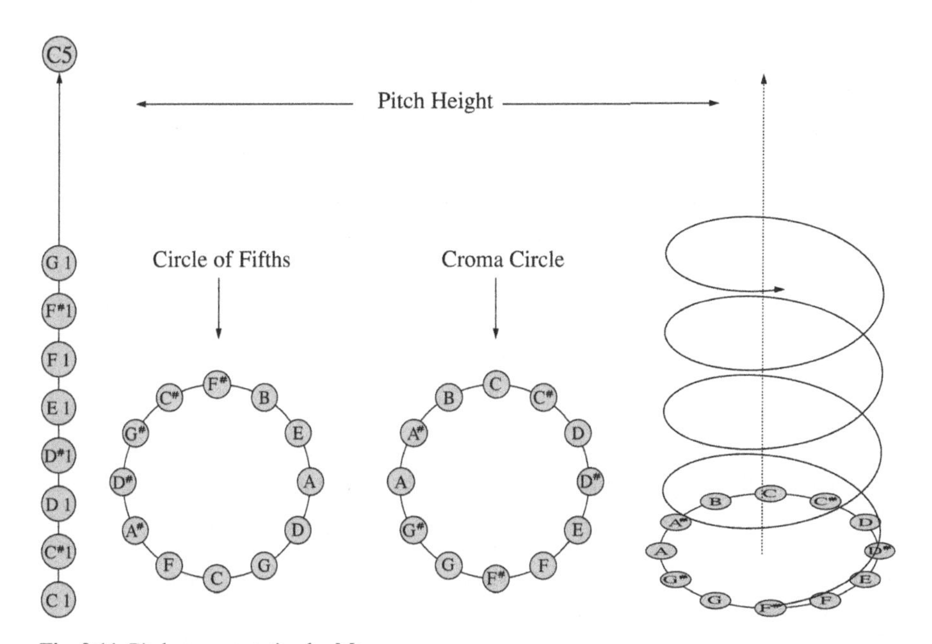

Fig. 9.11 Pitch representation by Mozer.

[14] Because CONCERT contains recurrent structures and the back-propagation algorithm is generally only used in the context of feedforward networks; see the modifications in [18, p. 7–8].

Based on a model by Roger Shepard [25], pitch is represented in a five-dimensional space (figure 9.11). The pitch class is determined by a position on the "chroma" circle and the absolute tone pitch is indicated by the position on a scale, the "pitch height." This particular form of pitch representation can be displayed within a three-dimensional helix-structure. Since tones of the same pitch class have the same value in one dimension, an adequate modeling of the perceptual similarity is possible. Two further dimensions indicate the pitches on the circle of fifths, which allows the modeling of additional aspects of auditory perception [18, p. 10]. To adapt this five-dimensional representation for different requirements, the components can be weighted by adjusting the diameters of the chroma circle and the circle of fifths. Chords are represented for each of the component pitches by the fundamental and the first four harmonics, mapped to the nearest pitch class of the chromatic scale. This representation is based on a proposal by Bernice Laden and Douglas H. Keefe [11] and modified by Mozer in some aspects [18, p. 15]. Additionally, chords that have a dominant or subdominant relationship to one another are rated as being higher in terms of their similarity. The representation of note durations allows for the formalization of the similarities between different durations by involving partial relations. Based on the encoding of each beat (quarters) as twelfths, note values are represented also in a multi-dimensional space (figure 9.12).

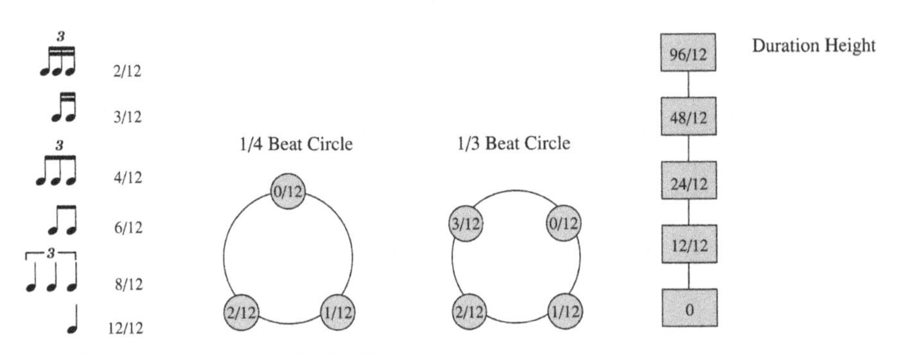

Fig. 9.12 Rhythmic representation by Mozer.

"Duration height" indicates the multiple of the quarter which is taken as a basic unit. Two additional values represent the duration as coordinates on a 1/3 and a 1/4 circle. The coordinate on the 1/n beat circles results from the duration after subtracting out the greatest integer multiple – as well as by performing a modulo operation. The following example is made with the durations 18/12 (quarters with held eights) and 15/12 (quarters with held 1/16): After performing the modulo operation for 18/12, a value of 2/12 on the 1/3 beat cycle and 0/12 on the 1/4 beat cycle results; for 15/12, the result is: 3/12 (1/3 beat) and 0/12 (1/4 beat). In this encoding the same value of both durations in relation to the 1/4 beat circle shows the similarities of both note values in terms of their composition of even note values (e.g. quarters and eights). The fact that both note values also have similar lengths is proof of their close positioning in "duration height."

CONCERT is initialized with a short sequence and produces musical material that is thereupon subject to a series of examinations. In the beginning, scales, *random walks*,[15] random walks with melodic jumps and simple form schemes are learned. Apart from the learning of formal components that possess a large context [18, p. 20], the generations of the system are satisfying. Consequently, CONCERT is used for the production of material in the style of simple folk melodies, soprano voices of different works of J.S. Bach, as well as waltzes of several composers. The productions of CONCERT prove to be superior to Markov chains of third order in a hearing trial. Even though CONCERT generates interesting musical segments on a small scale, the system, however, is lacking with regard to the structuring of longer musical sequences. In longer sequences these thematic relations are lost and the productions of the ANN tend to be arbitrary so that for the improvement of its efficiency, possible extensions of the system are presented [18, p. 26].

9.4.2 Boltzmann Machines and LSTMs

Matthew Bellgrad and Lawrence Peh Tsang [1] use a *Boltzmann machine* (BM) for the harmonization of given chorale melodies composed in the Baroque period. A Boltzmann machine is one variant of a Hopfield net. As in a Hopfield net, the BM consists in general of a single layer of completely connected neurons and serves to assign patterns as well. In order to prevent the net from being limited to local energy minima, in the processing of a cycle the BM may also take on states with a higher energy potential. This process is a form of *simulated annealing*, so called for the conceptual analogy of this algorithm to the hardening of metals. In a BM, the net energy is temporarily increased by means of a controllable temperature parameter that in each step may additionally modify the activation of a neuron through a probability-based value. For their model Bellgrad and Tsang use an extended version of a so-called *Effective Boltzmann Machine* (EBM) that is trained on the local contexts of a set of chorales. All chorales of the training set are transposed on a mutual key. The tone pitches are represented locally by 35 neurons that present pitches of an underlying scale. Three further neurons indicate formal segments. According to its use by Bellgrad and Tsang, an "event" is the interconnection of a number of neurons to form a chord in the shortest appearing duration; longer durations are produced by repetitions. The harmonic context is learned by means of nested Boltzmann machines that each examines the immediate harmonic neighborhood of the note currently sounding. In case the observed context is not contained in the same configuration in the training set, the model may also generate harmonizations with a differing number of voices, as shown in the upper part of figure 9.13. An extension of the system that comprises restricting rules avoids undesired voice crossings and guarantees the correct number of voices for each harmonization step. A harmoniza-

[15] Here: Melodies that change always one step at most in the respective tonal system per time unit.

tion of the soprano in the above figure after an examination by these constraints is illustrated in the lower part of figure 9.13.

Fig. 9.13 Example of an incorrect harmonization of a given soprano voice, due to lacking correspondence in the training set (top). Harmonization of the BM after implementation of constraints (bottom).

Jürgen Schmidhuber and Douglas Eck [23] use a *long short-term memory recurrent neural network* (LSTM)[16] for the production of melodic material over a given chord progression. A problem occurring in traditional network architectures lies in their treatment of context-sensitive material. Although common recurrent ANNs may consider a certain number of previously produced data, in a larger context, however, this type of network also shows some weaknesses, as Mozer states for his system: "While the local contours made sense, the pieces were not musically coherent, lacking thematic structure and having minimal phrase structure and rhythmic organization. It appears that the CONCERT architecture and training procedure do not scale well as the length of the pieces grows and as the amount of higher-order structure increases." [18, p. 26]. An LSTM consists of an input and output layer as well as a number of interconnected memory blocks. Each block contains a differing number of recurrent memory cells containing the information of previous states. Information is exchanged through *gates* that, according to the input and type of threshold function, may enable either the admission or transfer of information, pass on information after a delay or delete the content of the memory block. In most cases a conventional recurrent ANN may use ten to twelve previous steps for the context treatment, while an LSTM can treat a context of over 1000 timesteps.

Eck and Schmidhuber use a "local" representation of note values and rhythmic changes are produced by holds. For both the chord and melody material the tonal space of an octave is applied. In a first experiment, only chord progressions are learned using a form of twelve bar blues. Further, parallel to the chord sequence, melody lines are built based on a pentatonic scale. An evaluation carried out by a jazz musician finds clearly better results for these generations than for passages produced with a random walk method. Although a considerably larger context may

[16] Developed by Sepp Hochreiter and Jürgen Schmidhuber, cf. [5].

be treated by LSTMs in contrast to simple recurrent architectures, this model also reveals certain limitations in terms of the generation of larger musical sections.

In the context of algorithmic composition neural networks are, in most cases, not exclusively used for the generation of structure, but can be found, however, often as an extension of other procedures (for example, in the form of fitness raters in the field of genetic algorithms) or for the examination of single musical aspects. Alejandro Pazos, A. Santos del Riego, Julian Dorado and J.J. Romero-Cardalda [19] developed a system treating agogics in Western classical music. The musical information used is provided by means of rhythmic pulsations of a MIDI pedal made by the musician. These data make up the training material of the ANN that is used for the prediction of agogic variation.

Artificial neural networks may also be well applied in the field of classification where musical material is, among other things, examined in terms of the tonality of individual musical segments [26, 27] or stylistic distinctions [2, 8]. For these analyses, it is mainly Kohonen's self-organizing maps that are applied.

9.5 Synopsis

Artificial neural networks were developed – according to the scheme of information processing in the cortex – originally for tasks of image recognition and classification. In contrast to the biological model, the processing capacity of an artificial neural network is extremely low – however, these algorithms are also able to solve complex tasks within the original core applications, producing satisfactory results. There are only certain classes of neural networks that are suited to processing musical information; in some approaches, the architecture of network types is adapted in order to be able to optimally represent and produce specific properties of musical structure. ANNs are frequently applied in the framework of hybrid systems of algorithmic composition where they are used, for example, as fitness raters in the context of a genetic algorithm.

Besides the application of special network architecture, it is above all the representation of musical information that forms an essential aspect of musical structure genesis. The different possibilities of "distributed," "local," "absolute" and "relative" representation have, as already demonstrated by the works of Todd and Mozer, decisive effects on the processing of the musical inputs. Mozer follows an interesting approach in his system CONCERT. The complexity of musical information is taken into consideration by means of a multi-dimensional representation model that represents pitch, rhythm and harmony of the musical components under different aspects. In most cases, the output of neural networks consists of concrete note values; this system's output, however, consists of probabilities for the generation of particular note values. CONCERT and some other abovementioned systems as well, may generate satisfying results over short passages, but show weaknesses in the creation of larger context-dependent material. On the one hand, the reason for these difficulties lies in the application of the back-propagation algorithm used frequently in the training, which has problems in processing an exhaustive context. On the other hand, the modeling of large and context-dependent musical sections is a general problem for all methods of algorithmic composition that are not able to process information as a hierarchically ordered structure. For the treatment of material which is context-dependent over long passages, generative grammars are well suited. When knowledge about the domain to be modeled exists, rule-based systems are generally preferred. Moreover, the generations in a neural network often end up in stationary situations. Another disadvantage is the fact that for the production of longer musical segments, a great number of training cycles is in most cases required.

An essential point of criticism made by Mozer refers to the fact that generations of neural networks in algorithmic composition are mostly discussed uncritically. Either the material is not evaluated because a successful output is already rated as a success per se, or subjective criteria that do not enable an objective rating of the achieved results are used for an evaluation – this criticism points out a common lack in a number of publications in the field of algorithmic composition.

A great advantage of neural networks over Markov models and generative grammars, however, becomes evident in the production of smaller musical components. Both in generative grammars and also in Markov models, only transitions (e.g. of tone pitches) that are also explicitly contained in the corpus may be generated. Here, the neural network may produce surprising movements that nevertheless meet the requirements of the underlying corpus – this aspect of artificial neural networks may also represent an interesting motivation for their application in a framework of innovative compositional concepts.

References

1. Bellgrad MI, Tsang CP (1994) Harmonizing music the Boltzmann way. Connection Science, 6, 1994
2. Dannenberg RB, Thom B, Watson D (1997) A machine learning approach to musical style recognition. In: Proceedings of the 1997 International Computer Music Conference. International Computer Music Association, San Francisco
3. Hebb DO (1949) The organization of behavior. Wiley, New York
4. Hild H, Feulner J, Menzel W (1992) HARMONET: A neural net for harmonizing chorales in the style of J.S. Bach. Advances in Neural Information Processing Systems 4
5. Hochreiter S, Schmidhuber, J (1997) Long Short-Term Memory. Neural Computation, 9/8, 1997
6. Hopfield JJ (1982) Neural networks and physical systems with emergent collective computational abilities. Proceedings of the National Academy of Sciences of the United States of America, 79, 1982
7. Hopfield JJ (1984) Neurons with graded response have collective computational properties like those of two-state neurons. Proceedings of the National Academy of Sciences of the United States of America, 81, 1984
8. Kiernan F (2000) Score-based style recognition using artificial neural networks. In: Proceedings of the first International Conference on Music Information Retrieval (ISMIR 2000), Plymouth, Mass
9. Kohonen T (1972) Correlation Matrix Memories. IEEE Transactions on Computers C-21, 4, April 1972

10. Kohonen T (1982) Self-organized formation of topologically correct feature maps. Biological Cybernetics, 43, 1982
11. Laden B, Keefe DH (1989) The representation of pitch in a neural net model of chord classification. Computer Music Journal 13/4, 1989
12. Lämmel U, Cleve J (2001) Lehr- und Übungsbuch Künstliche Intelligenz. Fachbuchverlag Leipzig im Carl-Hanser-Verlag, München. ISBN 3-446-21421-6
13. Luger GF, Stubblefield W (1998) Artificial intelligence. Structures and strategies for complex problem solving, 3rd edn. Addison-Wesley Longman, Amsterdam. ISBN 0-805-31196-3
14. McCulloch W, Pitts W (1943) A logical calculus of the ideas immanent in nervous activity. Bulletin of Mathematical Biophysics, 5, 1943
15. McCulloch W (1945) A heterarchy of values determined by the topology of nervous nets. Bulletin of Mathematical Biophysics, 5, 1945
16. Minsky M, Papert S (1969) Perceptrons: An introduction to computational geometry. MIT Press, Cambridge, Mass
17. Miranda ER, Biles JA (eds) (2007) Evolutionary computer music. Springer, London. ISBN 978-1-84628-599-8
18. Mozer MC (1994) Neural network music composition by prediction: Exploring the benefits of psychoacoustic constraints and multiscale processing. Connection Science, 1994
19. Pazos A, del Riego SA, Dorado J, Romero-Cardalda JJ (1999) Connectionist system for music interpretation. In: International Joint Conference on Neural Networks (IJCNN'99), vol 6. IEEEE, New York, pp 4002–4005
20. Rosenblatt F (1962) Principles of neurodynamics. Spartan, New York
21. Rumelhart DE, Hinton GE, Williams R (1986) Learning internal representations by error propagation. In: Rumelhart D, McClelland JL (eds) (1986) Parallel distributed processing: Explorations in the microstructure of cognition, 1: Foundations, MIT Press, Cambridge, Mass
22. Rumelhart DE, Hinton GE, Williams R (1986) Learning internal representations by back-propagating errors. Nature, 323, 1986
23. Schmidhuber J, Eck D (2002) A first look at music composition using LSTM recurrent neural networks. Technical Report IDSIA-07-02, 2002
24. Schmidhuber J, Eck D (2005) Composing music with LSTM recurrent networks - Blues improvisation.
http://www.idsia.ch/ juergen/blues/index.html Cited 3 Nov 2005
25. Shepard RN (1982) Geometrical approximations to the structure of musical pitch. Psychological Review, 89/4, 1982
26. Tillmann B (1999) Connectionist simulation of tonal knowledge representation. Publications Journées d'Informatique Musicale, Issy-Les-Moulineaux, 1999
27. Tillmann B, Bharucha JJ, Bigand E (2000) Implicit learning of tonality: A self-organizing approach. Psychological Review, 107/4, 2000
28. Todd PM (1989) A connectionist approach to algorithmic composition. Computer Music Journal, 13/4, 1989
29. Werbos P (1974) Beyond regression: New tools for prediction and analysis in the behavioral sciences. PhD Thesis, Harvard University, 1974
30. Widrow B, Hoff ME (1960) Adaptive switching circuits. IRE Western Electric Show and Convention Record, 4, August 23, 1960

Chapter 10
Artificial Intelligence

The field of research on *artificial intelligence* (AI) combines a number of different disciplines dealing with the automatization of intelligent behavior. It is difficult to systematize the distinct approaches of AI for various reasons. Relevant methods are enumerated mostly in terms of their applicability to domain-specific tasks. So, many approaches are either integrated in a system of AI or treated as a separate discipline.

If a definition of intelligent behavior is used as a description model for relevant algorithms, a further problem arises due to the fact that the term "intelligence" is interpreted in various ways. In the following, some aspects of ELIZA,[1] one of the first programs for the simulation of intelligent behavior, are described in order to illustrate this. Joseph Weizenbaum's intention in designing this program was to create a mimicry of a conversational situation in a psychotherapeutic context. The "patient" (user) enters his or her questions in the program and the "psychotherapist" (ELIZA) answers them. The extract below illustrates a typical conversation between a user and ELIZA.

"Men are all alike.
IN WHAT WAY?
They're always bugging us about something or other.
CAN YOU THINK OF A SPECIFIC EXAMPLE?
Well, my boyfriend made me come here.
YOUR BOYFRIEND MADE YOU COME HERE.
He says I'm depressed much of the time.
I AM SORRY TO HEAR YOU ARE DEPRESSED.
It's true. I am unhappy [...]." [65, p. 36].

ELIZA uses the following principal strategies for the simulation of a conversation partner:

[1] Developed by Joseph Weizenbaum (1923–2008), mathematician and computer scientist; from 1964, teaching at the MIT, 1964–1967 work on language analysis programs. For ELIZA, see [65].

- Identification of *key words*[2] in the user's input text.
- Discovery of a minimal context of the particular key word.
- The choice of appropriate rules in order to transform the context of the input in the corresponding context of the output.
- Generation of uncontroversial and general responses in the absence of appropriate key words.

The efficiency of programs such as ELIZA in terms of simulating a human conversation partner is often examined by means of a test arrangement. In "Computing Machinery and Intelligence," [62] Alan Turing developed a virtual setup to evaluate a computer's intelligent behavior. In the beginning of his considerations, he designed a game in which a person is to guess the sex of two other conversation partners. A man, a woman and the person who carries out the test go into separate rooms; they can only communicate in writing. The testing person's task is now to find out the partner's sex through asking targeted questions. What makes this task even more difficult is that the man and the woman aim to convince the questioner that they are the opposite sex by giving deliberately true or false answers. If a computer then takes the part of one of the questioned persons, the essential question of the test (which has since become legendary) is: Who is the human and who is the machine?

Fig. 10.1 Alan Mathison Turing by Elliott & Fry. Quarter-plate glass negative, 29 March 1951. © National Portrait Gallery, London.

[2] Key words are words that may be included as carriers of meaning in a general context in the course of the further conversation, such as "boyfriend" or "depressed" in the abovementioned example.

It is evident that the Turing test could also be passed by a machine[3] that simulates intelligent behavior without, however, providing intelligent problem solving strategies. The rule system of ELIZA serves as an example to show that the "intelligent behavior" of a program is not to be achieved necessarily through AI specific algorithms of whatever type. In a tautological sense, all approaches that are used to solve AI specific tasks may be considered methods of AI. If the process of generating coherent musical structure is understood to be "compositional intelligence," then some of the already treated procedures of algorithmic composition may be ranked among the methods of AI as well. A prime example of an application of this "compositional intelligence" is surely Cope's system EMI, which, in the field of style imitation, performs a job comparable to that of a human expert.[4] In the framework of algorithmic composition, in some respects, this system also represents a parallel to ELIZA as can be demonstrated by means of a "musical Turing test."[5]

Generally, one may attempt to partially explain the field of AI-specific procedures by means of a definition of the term "intelligence." Etymologically, the term derives from Latin "intelligentia," formed by "inter legere," (i.e. to classify a term in the right category by critically considering its relevant properties). Due to the fact that this term is used in different humanistic and natural scientific disciplines, its definition must often be interpreted depending on the context. Recursive descriptions that, for example, explain intelligence through "intelligent behavior" of a biological species or a machine indeed exemplify the term, but leave an imprecise definition. The problem of a clear definition of the term is also precisely described by Turing in the context of his test: "I propose to consider the question, 'Can machines think?' This should begin with definitions of the meaning of the terms machine and think. The definitions might be framed so as to reflect, as far as possible, the normal use of the words, but this attitude is dangerous. If the meaning of the words machine and think are to be found by examining how they are commonly used, it is difficult to escape the conclusion that the meaning and the answer to the question, 'Can machines think?' is to be sought in statistical surveys such as a Gallup poll. But this is absurd [...]." [62]. Two definitions that represent a highly general possible explanation of the term are presented by scientists who, from early on, strove to establish a standard for the evaluation of "human" intelligence. The German psychologist and philosopher William Stern (1871–1938) coined the term "Intelligenzquotient" (intelligence quotient) and founded "differential psychology," a discipline which examines the differences between single persons in terms of their mental properties. Stern defines intelligence as the general ability of an individual to consciously adjust his or her thinking to new requirements; in this sense, intelligence represents a general adaptability to new tasks and conditions of life. According to the American

[3] In 1991, the Loebner Prize (100,000 USD) was created to award the first program to pass the Turing test to the fullest extent; up to today, no program has been able to completely meet these requirements. For the Loebner Prize, see [36, 55].

[4] Of course, EMI is a prominent procedure in algorithmic composition in the framework of AI; due to the fact that transition networks are an essential component of EMI, this type of network is treated in chapter 5.

[5] Cf. chapter "The Game," in [10, p. 13ff].

psychologist David Wechsler (1896–1981), who developed the concept for the first standardized intelligence test [63], intelligence is "the aggregate or global capacity of the individual to act purposefully, to think rationally and to deal effectively with his environment." [63, p. 3]. Based on general definitions, different descriptions of artificial intelligence can be made, such as the suggestion of Avron Barr and Edward A. Feigenbaum: "Artificial Intelligence (AI) is the part of computer science concerned with designing intelligent computer systems; that is, systems that exhibit the characteristics we associate with intelligence in human behavior – understanding language, learning, reasoning, solving problems and so on." [17, p. 3]. This definition shows two possible manifestations of artificial intelligence implicitly. In one case, AI programs are developed for problems that require "intelligent" behavior in order to be solved. In the other case, the aim of the program is to imitate "intelligent" behavior, as done, for example, by ELIZA through the simulation of a conversation partner. This second aspect can also be found explicitly expressed in a definition of AI by the computer scientist John McCarthy on the occasion of the Dartmouth Conference of 1956, an event that is often considered the beginning of the research field of AI: "[...] making a machine behave in ways that would be called intelligent if a human were so behaving." Because algorithmic composition in the context of artificial intelligence searches for "intelligent" problem solving strategies for musical tasks, a definition in terms of imitation can be left unconsidered for this field. Even though processes that are modeled on human compositional strategies[6] are integrated in systems of algorithmic composition, such can be naturally understood as an extended problem solving approach rather than an imitation of intelligent behavior.

10.1 Algorithmic Composition in AI

Algorithmic composition in the framework of artificial intelligence may be presented from different points. Based on exemplary publications in this field, there are methods that are either assigned to AI or described outside this area.[7] Furthermore, in algorithmic composition several methods of AI are not exclusively used for the generation of musical structure, but represent components of comprehensive systems. This applies above all to principles of knowledge processing in AI. In this context, the state space search is relevant to many data intensive methods, and therefore is also part of numerous algorithmic composition procedures, without, however, having to represent an independent composition technique. Based on the treatment of several introductory publications[8], this chapter attempts to estab-

[6] E.g. Variations in [28] and [29], see chapter 7.

[7] So, e.g. in the anthology "Understanding Music with AI" [6] a series of approaches can be found that in the framework of this work are subsumed under separate methods, such as e.g. the works of Kippen and Bel (see chapter 4) and others.

[8] Feigenbaum [17], [18], [19], Russell [52], Winston [68], Kreutzer [32], Goerz [23], Luger [35], Lämmel [33].

lish a system of AI which contains all those approaches that are important for the generation of musical structure.

In the creation of such a system it turns out to be difficult to find a uniform classification for the different approaches. Although particular methods such as first-order predicate logic, expert systems, agents, etc. are treated more or less in detail in some works, for the most part different approaches are taken regarding their classification. In the following, only a few examples should be mentioned representatively: Luger and Stubblefield treat neural networks as a sub category of machine learning. Görz describes them in a separate section, whereas, for example, non-monotonic reasoning represents a category of machine learning. This method, on the other hand, can be found in Luger and Stubblefield in the category "reasoning with uncertain or incomplete information" as a sub category of "representations for knowledge-based problem solving," etc. Due to this different classification, for the representation of algorithmic composition in the context of AI, a structure as simple as possible is developed (see figure 10.2), in which some techniques may represent components of hybrid systems without, however, having to be explicitly treated as separate procedures. These include, for example, the way knowledge is represented, the possibilities of state space search or the application of logical operations in the context of a program sequence. So, for example, the software OpenMusic of IRCAM (Institut de Recherche et Coordination Acoustique/Musique) has a number of libraries that provide different procedures such as logical operations or rewriting systems. In some works of Phon-Amnuaisuk, Alan Smaill and Geraint Wiggins [44, 45], the encoding of musical material, among other things, is carried out in the logical programming language Prolog; Christopher Fry [22] and Kemal Ebcioglu [14] are developing their own representation languages within their software systems.

In algorithmic composition, it is not always fully possible to assign the mentioned works to the context of a particular method because several developments in this field are set out as hybrid systems and therefore apply a number of differing procedures for musical structure genesis.

In general, systems of algorithmic composition that generate musical structure based on domain-specific knowledge require that this information be represented in one way or another. Knowledge representation consists basically of the possibility of formally storing knowledge in an information system and further drawing conclusions on the basis of this information with the help of various procedures. As for algorithmic composition, in the simplest case, a number of musically interpretable data may be collected out of which parts are selected randomly for the composition process. In order to establish relationships between the separate data and to enable conclusions that can be arrived at when examining information from different perspectives, a number of methods for knowledge representation, such as experts systems, are used in algorithmic composition.

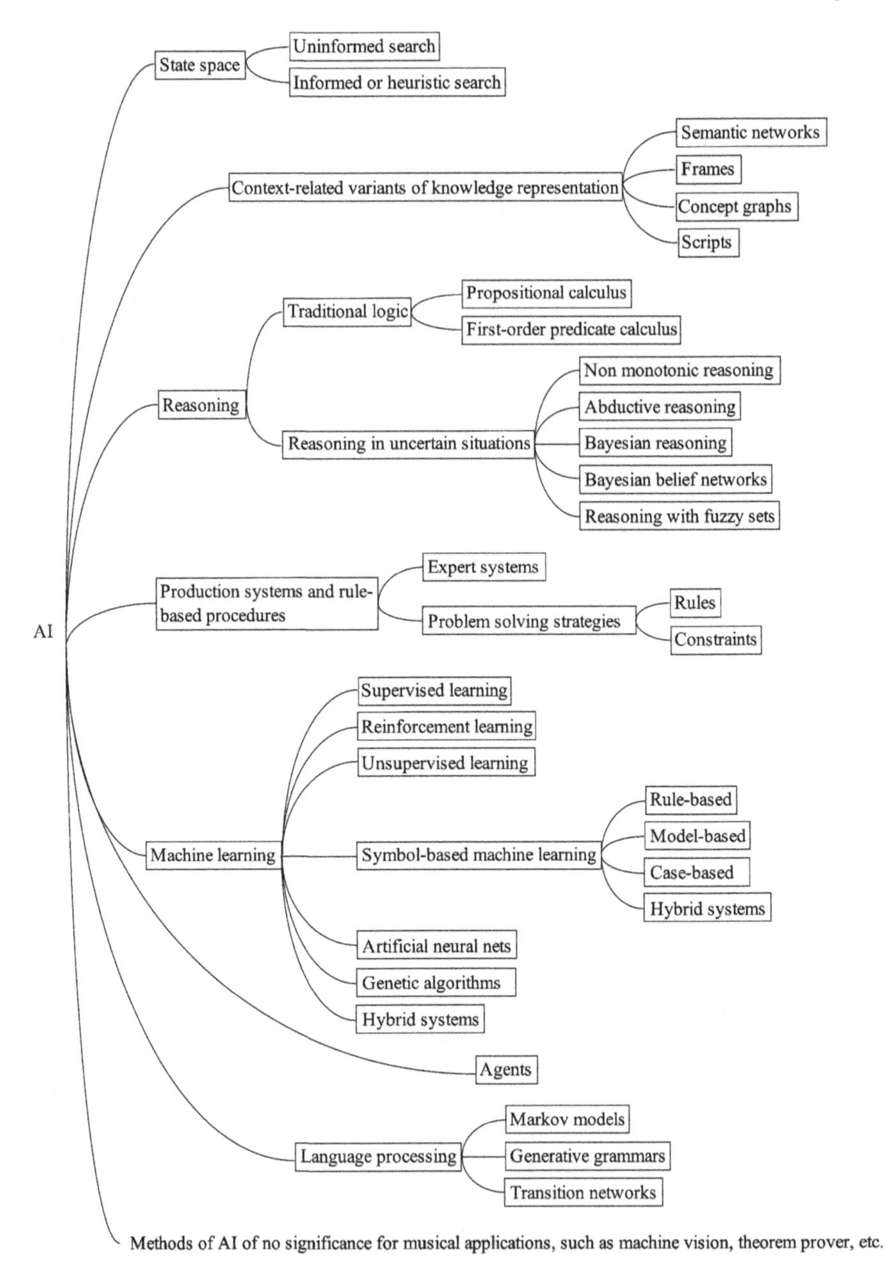

Fig. 10.2 Algorithmic composition in the framework of artificial intelligence.

10.1.1 State Space

However the representation of knowledge is formalized in these systems, it brings about the definition of a *state/search space* that represents all possible events of the respective discourse field. For the representation of the state space, a graph notation may be used in various ways. If particular directions are assigned to the edges, it is referred to as a *directed graph*. A *root graph* begins precisely at one node. Within a *tree* the nodes of a graph are connected in pairs solely by one edge. Figure 10.3 illustrates the state space by means of the possible movements of the knight on a chess board.

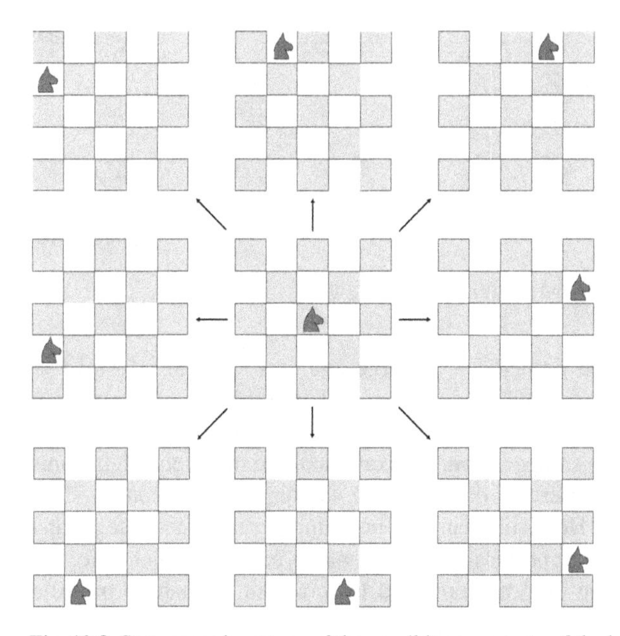

Fig. 10.3 State space by means of the possible movements of the knight on a chess board.

In the so-called *uninformed search*, the state space is traversed fully. *Forward chaining* is a reasoning method that starts searching from the initial nodes to the goal nodes, i.e. from facts to conclusions. This strategy is also referred to as a *data-driven search*. In *backward chaining*, the state space graph is searched starting from the goal nodes to the initial nodes. This method is also called a *goal-driven search*. Depending on the task, the search in forward and backward chaining results in state spaces of different size. As an example, the search for an ancestor within ten generations could be of use here. In forward chaining, two parents must always be taken into consideration; therefore 2^{10} states result. If every pair of parents is assumed to have three children, in backward chaining, however, a considerably larger state space of 3^{10} states results [35, p. 117].

Irrespective of the direction of the search, the state space may be searched in two different ways. In *depth-first search*, a node of the same level is only searched when the node that had been searched before has been traversed in all deeper levels. *Breadth-first search* is an algorithm that searches every level in the search graph completely; only after this has finished the next level is explored, as can be seen in figure 10.4.

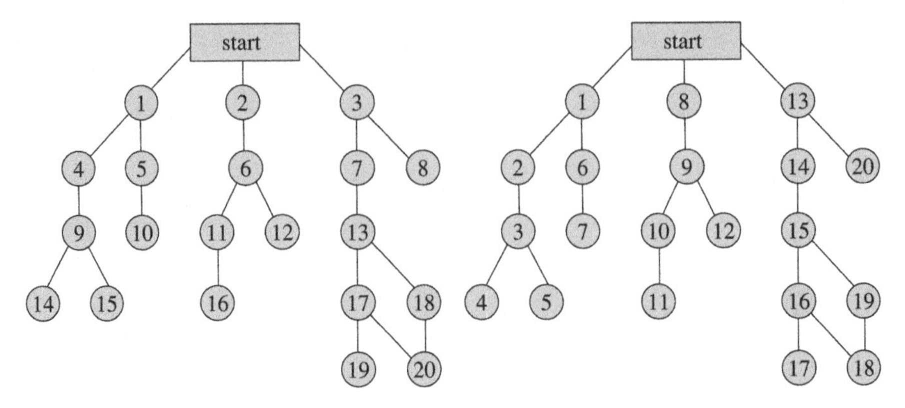

Fig. 10.4 Breadth-first search (left) and depth-first search (right).

If the state space must be limited due to large amounts of data, informed or heuristic search strategies are applied that integrate probability-based estimations of the achievement of a particular goal in the problem solving process. In algorithmic composition, heuristic search strategies are, for example, used in genetic algorithms in which large amounts of generated data must be examined in the course of an automatic fitness evaluation. Heuristics can also be found in real-time systems that are developed for interactive improvisation or automatic accompaniment. So, the live recordings of a musician may be compared to scores in a database in order to generate appropriate accompanying voices. An example of an accompaniment system of this kind can be found in the work of Roger Dannenberg [13]. In this system, heuristics are applied in the comparison of patterns of input values that try to find the most similar structures possible in the stored data. In an application of Randall Spangler [56], musical material is analyzed in order to extract rules of harmony; these rules are then applied to generate new harmonizations in response to a melody input in real-time by heuristic methods.

In reference to a general definition of the term "heuristic,"[9] these techniques are part of nearly every algorithmic composition system. In a stricter sense, heuristics are often referred to as methods for limiting state space search. Different algorithms provide, in this context, solutions for a state space search which is optimized in regard to the data to be examined. The solutions of a heuristic state space search, however, render only uncertain results, in contrast to a complete traversing of the

[9] E.g. "involving or serving as an aid to learning, discovery, or problem-solving by experimental and especially trial-and-error methods" [16].

state space, also called a "brute-force search." The more effort required for its calculation, the higher the probability of success of a heuristic algorithm. A simple example of a problem for which heuristics do not always find an optimal solution is the *Traveling Salesman Problem*. The problem statement is to find the shortest route between a number of cities. Along the route, the salesman must visit every city exactly once and then return to the starting point again. The solution seems simple, but even with a relatively small number of cities an exhaustive traverse of the state space requires enormous calculation efforts due to the large number of possible combinations, commonly called combinatorial explosion.[10] One of the simplest heuristics that, in contrast, requires only little expenditure of calculation is the *Nearest Neighbor Heuristic*. This method continues searching only from the best nearest node. By means of breadth-first search, the search is continued only from the optimal node of the first level; no other paths are taken into consideration. The same principle is applied to every further level. If, for example, the nearest neighbor heuristic is used at this point to solve the Traveling Salesman Problem, then additional solutions may also be found that do not describe the optimal route between the separate cities, as shown in figure 10.5 (right).

Fig. 10.5 Two solutions of a Nearest Neighbor Heuristic for the Traveling Salesman Problem

10.1.2 Context-Dependent Variants of Knowledge Representation

Semantic networks represent different types of objects and classes within a graph. In that representation, the nodes stand for objects, classes and subclasses; the edges are labeled to mark the relations between the separate nodes. Figure 10.6 illustrates through a semantic network that all violinists are musicians, that Sarah is a particular example of a violinist – and is consequently also a musician – and also, that Sarah owns a Stradivari, which again is a special example of the class "violin."

Based on the observation that humans describe objects of a class by means of particular characteristics, knowledge may also be represented in the form of so-called *frames* that produce a standardized context for stereotypical situations. Frames provide the framework for a particular context and may be represented within a seman-

[10] For example, for 4 cities (3*2*1) = 6 possible routes, for 8 cities = 5040, and for 12 cities already more than 479 million.

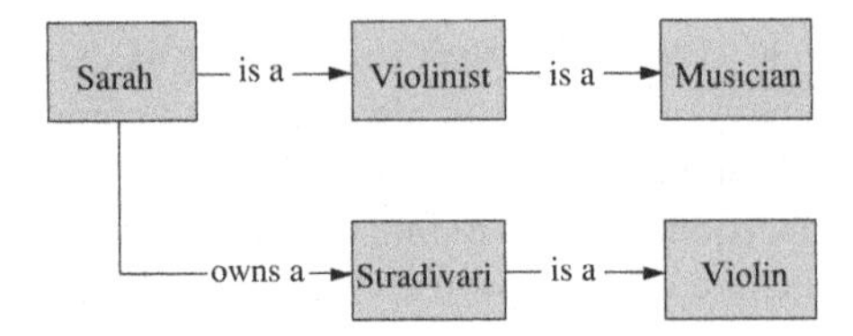

Fig. 10.6 Semantic network.

tic network. Consequently, a car has, in most cases, four tires; in a kitchen one can find an oven of some kind, etc.

In contrast to frames, *scripts* structure stereotypical sequences within a context. An example of a script could be the process of shopping: First, one enters a shop, then the goods are selected; finally one must inevitably go to the cash register. By embedding pieces of information in such a standardized context, multiple semantic meanings may be clarified and also, information that is not contained in the input may be inferred from the context. On a shopping trip, for example, at the stage "payment," it may be assumed that this happens at a cash register, a shop assistant is involved, etc.

An additional form of knowledge representation in which elements of a discourse field are connected by particular relations, is *concept graphs*. Within the graph representation, the nodes may either represent elements or relations. The elements of a concept graph are referred to as concepts (the boxes in figure 10.7) and may illustrate concrete or abstract objects, or also, particular properties of an object. By means of *concept relations* (circles in figure 10.7), the concepts are related to each other in different ways. The network of relationships is represented by a bipartite graph,[11] in which the concepts may only be connected through concept relations. As an example, figure 10.7 shows the representation of the sentence "a child eats a pudding with a plastic spoon" by means of a concept graph. Here, AGNT refers to the *agent*, i.e. the doer of the action; PTNT is the object affected by the process, the *patient*; INST is the instrument used to carry out the action; MATR stands for *material* and represents the relation between the instrument and the material.

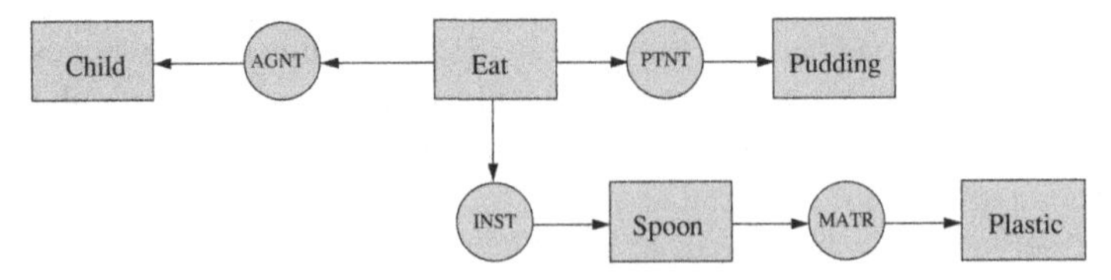

Fig. 10.7 Concept graph.

[11] A graph is called bipartite when its nodes may be divided into two disjoint sets (bipartition) so that there are no edges between the nodes within a set (meaning the concepts and the concept relations).

Semantic networks or concept graphs find their parallels in different systems of algorithmic composition that process information within classes, objects and their mutual relations. If basic conditions are formulated for particular musical classes, this approach can be expressed in AI through the concept of frames. Scripts are models for the chronological organization of exemplary sequences of musical events.

10.1.3 Reasoning

A number of systems of algorithmic composition make use of propositional calculus and first-order predicate calculus for the encoding and derivation of musical information.[12] Extensions of classical logic enable *reasoning in uncertain situations*. Classical logic has the property of *monotonicity of entailment*, which states that new knowledge may indeed be obtained through logical operations; already proven facts, on the other hand, cannot be revised any more.

Methods of *non-monotonic reasoning*, in contrast, replace axioms of a calculus by assumptions that can be further replaced by new assumptions. One form of non-monotonic reasoning is *abductive reasoning* in which unknown reasons are inferred from known facts by forming hypotheses. In this sense, for example, smoke may point to fire.[13] Abductive reasoning is also referred to as "inference to the best possible explanation." Due to the fact that a conclusion derived by inference is not necessarily ascribable to an underlying premise or reason, abduction is not a correct rule of reasoning in the sense of classical logic. Just like abduction, induction (reasoning from the special to the general) may only make a proposition of an assertion. Although these methods of reasoning use proposition instead of proof, they are nevertheless often used in adaptive systems that are able to renew already made assumptions along with increasing information. These forms of knowledge acquisition are also basic means of natural scientific research whose findings are subject to permanent changes as well. In addition, case-based expert systems that are often applied in algorithmic composition must be seen in the context of these techniques.

In the sense of traditional logic, *fuzzy sets* extend the "law of the excluded middle" in terms of the membership to a particular set. In a classical two-valued logic, a proposition is either true or false. In a multi-valued logic, such as fuzzy logic, the truth value – a value that indicates to what extent a proposition belongs to a particular set – may express itself in an interval between 0 and 1. In the context of fuzzy sets, traditional rules of reasoning and memberships to a set are discrete operations that may be extended by means of fuzzy sets in the sense of smooth state transitions. An example of fuzzy sets that are used for the categorization of sizes is shown in

[12] Apart from the already treated approaches, an early algorithmic composition system of IRCAM named CARLA must also be mentioned here (CARLA: Composition Assistée par Représentation Logique et Apprentissage); this system enables the encoding of knowledge on musical objects by means of the predicate calculus; cf. [11].

[13] Cf. the syllogism in Buddhist logic in chapter 2.

figure 10.8. In this case, for example, one and the same T-shirt can be of the size "small" or "medium."

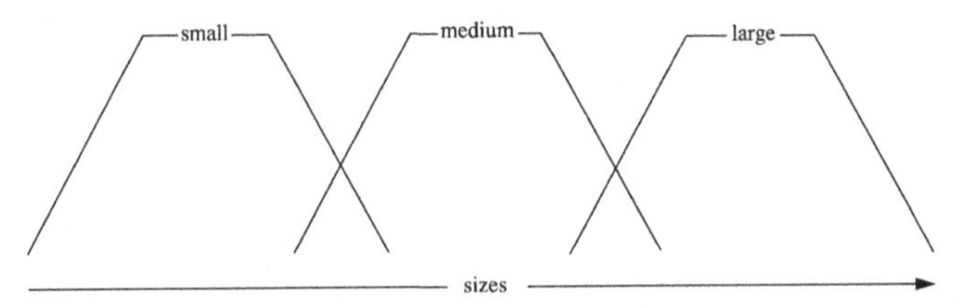

Fig. 10.8 Fuzzy sets by means of size.

On the one hand, fuzzy logic in algorithmic composition makes the definition of imprecise areas of musical parameters possible; on the other hand, the application of rules may be variably weighted by means of fuzzy sets that represent differing musical solutions. As a model, fuzzy logic is well suited to the representation of musical value ranges because musical components cannot always be assigned clearly to a particular category. In this sense, for example, dynamic indications must be seen relatively, chords may have different functions in a harmonic context, and so on. Manzolli and Moroni use in their system Vox Populi[14] fuzzy sets for the scope of voice ranges. Peter Elsea[15] shows different applications of fuzzy logic in terms of musical parameters and the selection of chord inversions. The membership of single notes to different scales, as well as dynamic markings, are mentioned in this work to serve as examples.

In the traditional probability calculus, the probability p of an event x is calculated on the basis of a given hypothesis. *Bayesian reasoning* infers the probability of a proposed hypothesis based on an observed event. Different hypotheses are assumed to have different probabilities; samples are taken then that correct the original assumptions. The revision of the evaluations shows the learning effect of Bayesian reasoning. In a *Bayesian network* (also: *belief network*) the approach of Bayesian reasoning is represented in a directed *acyclic graph* (no cyclic structures occur). Bayesian reasoning is not explicitly a procedure of algorithmic composition; however, it is frequently applied in the field of musical analysis [9, 46].

10.1.4 Production Systems and Rule-Based Systems

A *production system* is a model of a rule-based system that employs pattern matching in the problem solving process. In most cases, a production system consists of a set of rules (called productions), a working memory (or database), and a long-term memory. The system works by running repeatedly through a cycle which is made

[14] [38], see chapter 7.

[15] In: "Fuzzy Logic and Musical Decisions" [15].

up of three stages: recognize, resolve, and act. Production rules consist of preconditions and actions, the preconditions determining which action is executed at what time. The working memory has a description of the environment of the production system, meaning the facts to which the rules of the production system may refer. The rules are matched against the current state of the working memory, which results in the *conflict set*, a collection of rules following from the previous matching algorithm. The conflict set represents the preconditions that the rules may apply. The actions determined by the preconditions are executed until no precondition is left to be applied to the content of the working memory – a process referred to as the *recognize-act cycle*. In this type of system, either a data-oriented or a goal-oriented problem solving process may be carried out. Forward chaining in a production system works similarly to the recognize-act cycle – by means of rules and data, the system searches the path to the given target precondition. In a system that uses backward chaining instead of the preconditions, the actions are matched against the working memory and the preconditions are included in the memory. This new state is again compared to other action parts of the production rules, etc. The search ends when the preconditions of all production rules have been applied.

An *expert system* that, in its different forms, is structured upon the concept of a production system uses subject-specific knowledge for solving and evaluating given facts. In a rule-based expert system, the knowledge base consists of a set of production rules that are divided up into "if-then" conditions; case-specific data is stored in the working memory. In most cases, rule-based expert systems apply forward chaining; however, backward chaining is also possible in the context of hypothesis examination.

Rules and *constraints* are knowledge-based problem solving strategies, meaning that there must be knowledge about the principles of the structure to be generated. A method of machine learning may be used here to gain knowledge about a problem that is further encoded in rules and constraints. In general, rules are formulated as "if-then" conditions. Constraints are conditions that limit the state space and may therefore give value ranges for variables that, for example, in algorithmic composition, limit the scope of musical parameters. Rules and constraints are often applied within a two-level procedure that first produces musical structure by means of rules that is further examined by constraints. As an example, a proceeding could demand tonics after a dominant function (the rule); the possible chord combinations are then examined in regard to the occurrence of parallels of perfect consonances (limiting condition in the sense of a constraint). In the literature, approaches in algorithmic composition that work with rules and/or constraints are often classified under the one or the other term.[16] Due to the fact that most approaches for algorithmic composition in this field use both rules and constraints, the works treated here are subsumed under the term "rule-based." This definition also results from the possibility of labeling constraints as a special class of rules that exclude objects of a discourse range from further processing within a system.[17]

[16] "The first two chords violate the *parallel fifth* rule [. . .]" [42].

[17] Examples for the use of constraints for the generations of chord progressions can be found in [47] and [24].

If the terms "rule" and "constraint" are applied in the sense of aesthetic categories irrespective of their informational importance, a distinction between the terms is, however, both interesting and meaningful. Early systems in the field of algorithmic composition often use *Generate And Test* (*GAT*) methods. The principle here is the generation of a number of values that in a further step are examined in terms of their musical usability.

The first example of a computer-generated composition produced by a GAT method is the "Illiac Suite" of Lejaren Hiller and Leonard Isaacson.[18] Regardless of the application of Markov chains, random values are generated that in further steps are examined for their musical usability by limiting rules. In the generation of the "Illiac Suite," the constraint is, in its notional meaning, also an aesthetic concept of the compositional process: "Moreover, the process of musical composition itself has been described by many writers as involving a series of choices of musical elements from an essentially limitless variety of musical raw materials. The act of composing, therefore, can be thought of as the extraction of order out of a chaotic environment. [...] In general, the more freedom we have to choose notes or intervals to build up a musical composition, the greater number of possible compositions we might be able to produce in the particular musical medium being utilized. Controversially, if the number of restrictions is great, the number of possible compositions is reduced. As an initial working premise, it is suggested that successful musical compositions as a general rule are either centrally placed between the extremes of order and disorder, and that stylistic differences depend to a considerable extent upon fluctuation relative to these two poles." [25, p. 9]. The analogy of the compositional process to a directed process of selection is an interesting aesthetical concept, but, however, does not correspond to the reality of compositional practice. Even though an examination of musical generations is carried out in different stages of the compositional process, the initial material already represents a limited number of applicable structural components and is not a priori the total of all possibilities.

Regardless of its aesthetical implication, this approach must certainly also be seen as the pragmatic approach of Hiller and Isaacson in which for the realization of the "Illiac Suite," elements of a discourse field are continuously eliminated according to a rule system in order to achieve correct results. Also, Hiller and Isaacson's hypothesis regarding "successful composition" which is located somewhere between the two poles of order and disorder, does not claim universal validity; however, due to its conception it may be easily formalized in the framework of a system of algorithmic composition. In the "Illiac Suite," Hiller and Isaacson work with different "experiments" that treat separate musical tasks. In experiments 1 and 2, counterpoint techniques modeled on the concepts of Josquin de Près and Giovanni Pierluigi da Palestrina are used for the generation of musical material. A less restrictive rule system is implemented in experiment 3 for the generation of "experimental music." Experiment 4 is an application of Markov chains of different order. The basic mode of operation in experiments 1 to 3 is well illustrated by means of counterpoint rules in the following text passage: "The music so produced was arranged

[18] Cf. [25], for the "Illiac-Suite," also see chapters 2 and 3.

to start with random white-note music [meaning here the white keys on a piano] and then by the successive addition of counterpoint rules was forced to progress gradually to more and more cantus firmus settings. We thought this procedure would provide an example of how order or redundancy might be brought into a musical texture." [25, p. 11]. Figure 10.9 shows the main routine for chromatic music as a part of the generation process of "experimental music" (experiment 3). The generation of the rhythmic structure is done by the generation of random binary numbers representing holds or rests, depending on playing instructions generated by an additional routine. Figure 10.10 shows the basic rhythmic schemes and their coding for 4/8 meter.

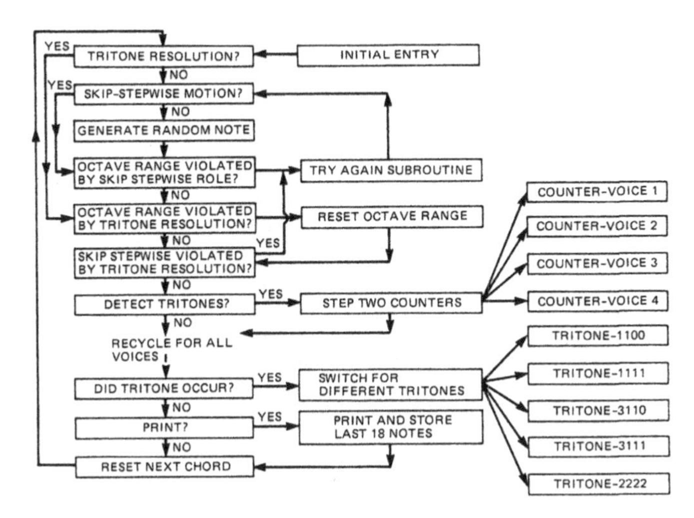

Fig. 10.9 Main routine for chromatic music in experiment 3 [25, p. 17]. © 1993 Massachusetts Institute of Technology. By permission of The MIT Press.

The "Illiac Suite" marks the beginning of a number of computer generated compositions. In the context of these works, the first computer music languages are subsequently being developed. In this field as well, Hiller is doing pioneer work together with Robert Baker. Their computer music language Musicomp, which basically consists of several small programs that can be combined, was designed in the early 1960s.[19]

A complex application of rule-based systems in the context of algorithmic composition can be found in Kemal Ebcioglu's system CHORAL, which harmonizes melodies in the chorale style of J.S. Bach [14]. For the software, a proprietary programming language is being developed; besides rules and constraints, heuristic procedures are also applied. Ebcioglu's programming language BSL (Backtrack-

[19] A well-known composition that was realized with Musicomp is the "Computer-Cantata" by Hiller and Baker (1963). For Musicomp, see [26, p. 37ff], for "Computer Cantata" [26, p. 50ff], for further works developed after the "Illiac Suite," see [26, p. 36].

Decimal number	Binary number	Rhythms: Closed	Rhythms: Open
0	0000		
1	0001		
2	0010		
3	0011		
4	0100		
5	0101		
6	0110		
7	0111		
8	1000		
9	1001		
10	1010		
11	1011		
12	1100		
13	1101		
14	1110		
15	1111		

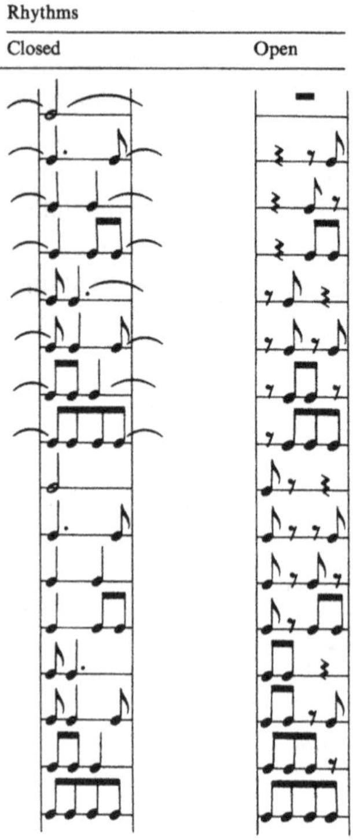

Fig. 10.10 Coding of rhythmic patterns [25, p. 19]. © 1993 Massachusetts Institute of Technology. By permission of The MIT Press.

ing Specification Language) encodes the domain-specific knowledge in a formalism similar to first-order predicate calculus and, due to its application-specific design, is capable of faster processing speeds than Prolog, Lisp and other similar languages – an advantage that, however, because of the processing speeds of today's computers, is rather of historical importance.[20] The harmonization in CHORAL takes place in a multi-level process that was designed according to Schenker's hierarchical system. Based on a skeleton of fermatas and rhythmless chords, the single voice leadings are generated in further steps. This architecture illustrates another type of rule-based systems in algorithmic composition that is applied frequently. The generation process is motivated less by the GAT method which examines randomly generated values with a number of functions in terms of their musical us-

[20] Prominent systems of algorithmic composition such as CommonMusic (CM) by Rick Taube or OpenMusic (OM) by IRCAM are recently developed Lisp-based systems that are constantly being further developed. CM also enables the processing of the stringent Lisp dialect "Scheme." Taube's book "Notes from the Meta Level" [58] includes a detailed description of CM for different applications of algorithmic composition. A good introduction to the concepts of OM can be found in [39, p. 967ff]. A number of articles treating algorithmic composition in the framework of OM can be found in "The OM Composers Book - Vol. 1." [1]; Vol. 2 to be published in 2008.

ability. Rather, a rule-based system is applied here hierarchically, which structures the composition step by step.[21] This structuring happens in Ebcioglu's system in separate procedures that build on one another: According to the given melody, the "chord skeleton view" generates several rhythmless chord progressions on the basis of rules and constraints; the heuristics placed in this view prefer cadences in the style of Bach. The "fill in view" that in turn is divided up into a number of procedures,[22] generates possible note values based on these progressions. A number of rules and constraints control the selection of notes in terms of their function as passing tones, suspensions, ornamentations, etc. The notes receive these functions as a result of their position, harmonic function, and meaning in a Schenkerian context that is given by the "Schenkerian analysis view."

If the results of the harmonization do not correspond to the constraints and heuristics, backtracking is applied. Figure 10.11 illustrates a chorale melody harmonized by CHORAL."

An additional approach that also generates chorale harmonization by means of rule-based strategies is being developed by Somnuk Phon-Amnuaisuk, Alan Smaill and Geraint Wiggins [44]. An inference module written in Prolog accesses a data base in which domain-specific knowledge is encoded in the form of musical structures and operations on these structures. An example of this architecture is illustrated in figure 10.12. As in Ebcioglu's work, the generation is carried out within a hierarchical model in which, due to a transparent program structure, control structures may be variably designed and the system may be extended for different applications.

Figure 10.13 provides an example of two harmonization versions of the same chorale line in different rhythmic densities.

Rule-based systems in algorithmic composition use different strategies for the representation and processing of music-specific knowledge. Apart from the rule (as an "if then" condition), constraints and the logical clause, generative grammars offer an additional possibility for encoding knowledge. Charles Ames and Michael Domino [2] use this formalism as a part of their Cybernetic Composer, which produces musical pieces of popular genres. Like the abovementioned systems, this software also uses constraints and backtracking in addition to other methods.

Most approaches to algorithmic composition generate musical structures in form of a score, i.e. requiring interpretation in terms of agogics and dynamics by performers. By contrast, some systems model performance practice of existing scores, thus studying the often implicit notions of musical interpretation.

Margret Johnson [30] developed a system that processes an input consisting of fugues from the Well-Tempered Clavier and outputs rules regarding the fugues' tempo, articulation and application of some ornaments. The used instrument is assumed to be a cembalo; due to this, dynamical structure may be disregarded. The in-

[21] Another interesting application of a hierarchical rule system can be found in [34]. David Levitt designed a system that generates a composition on the basis of simple basic conditions by using constraints of increasing specialization.

[22] Like, for example, "time-slice view" and "melodic-string view" which observe both vertically harmonic and horizontally melodic aspects of the harmonization in terms of a so-called "multiple viewpoint system"; cf. [14, p. 392ff].

Fig. 10.11 Chorale harmonization by Kemal Ebcioglu's system [14, p. 396, 397]. © 1993 Massachusetts Institute of Technology. By permission of The MIT Press.

terpretation rules are based on the expertise of two professional performers. Knowledge as to the correct interpretation is, on the one hand, determined by the rules,[23] and on the other hand by procedures that indicate their preferences for the application of the rules. The system examines the fugue by means of a tree structure in which the rules are encoded, and gives instructions for their interpretation, as partly shown in table 10.1. This approach therefore enables the recognition of motifs that are consequently, however, always interpreted in the same way. Although it promotes the recognition of identical motifs, this approach is not able to treat

[23] E.g. "If there are four 16th notes and a leap of greater than a third between the first and the second, then play the first 16th note staccato and slur the last three 16th notes" [30, p. 45].

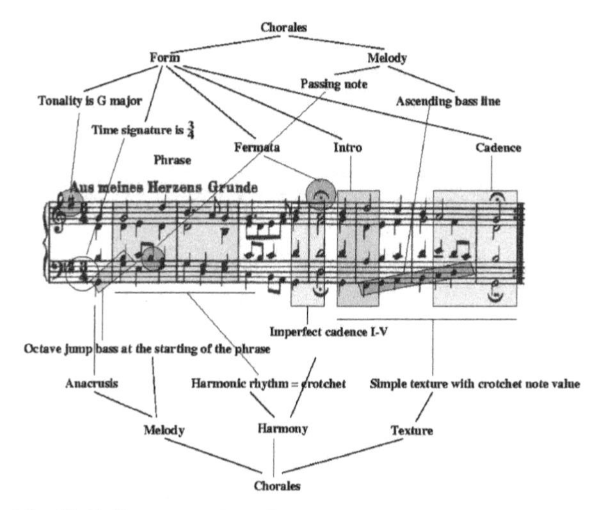

Fig. 10.12 Representation of musical structures with Phon-Amnuaisuk et al. Kindly provided by Somnuk Phon-Amnuaisuk.

Fig. 10.13 Two harmonization versions of the same chorale with Phon-Amnuaisuk et al.

their musical context, because the repetition of a motif may be musically motivated in different ways – a fact that in a successful "human" interpretation also becomes evident in different interpretations of the same motif. Anders Friberg, Vittorio Colombo, Lars Frydén and Johan Sundberg [21] developed Director Musices, a system for the musical interpretation of MIDI files. In this system, rules are formulated within categories on different hierarchical levels. In this way, a rule may affect an entire segment and also a particular event in this segment at the same time. The corpus of rules is particularly comprehensive and consists of the following categories: "Grouping rules" mark boundaries between note groups, "differentiation rules" produce contrasts between categories (e.g. loud – silent, short – long), and "ensemble rules" enable interactions between musicians in an ensemble and relate them to general specifications (e.g. interpretation with a swing feel). Combinations of different

Bar	Beat	Message
		Base tempo: M. M. quarter note = 80; fastest note value = 16th.
		Opening: Slur the two 16th notes together.
1	2	Subject: Play both eighth notes staccato.
1	3	Subject: Prolong the downbeat note slightly.
1	3	Subject: Slur the two 16th notes together.
1	4	Subject: Play both eighth notes staccato.

Table 10.1 Output of Johnson's system for the interpretation of Bach fugues.

rules are used to produce interpretations of a particular "emotional quality." The degree of application of the rules (like the higher, the louder) can be modified during playback in real-time by a user.

Some systems of algorithmic composition and analysis aim to model or recognize the "emotional content" of musical material. In these systems particular musical information is classified, for the most part, under a certain "emotional aspect."

Douglas Riecken's [50] system WOLFGANG is based on inductive machine learning in a knowledge-based system. In order to generate compositions in a characteristic style, four distinct emotions are assigned to the musical units in different shares. These components are combined within a network[24] and should make a genuine compositional process comprehensible by means of concerted "emotional qualities."

The Kansei Music System by Haruhiro Katayose and Seiji Inochuki [31] attempts to analyze musical material by means of emotional and extra-musical labeling: "We have assumed that sentiment is inspired from recognized musical primitives, such as melody, chord progression, rhythm, timbre, and their combinations. We therefore designed an experiment to extract sentiments using music analysis rules that extract musical primitives from transcribed notes and rules that describe the relation between musical primitives and sentiments." [31, p. 74–75].

In these approaches consideration must be given to the fact that the "emotional content" of a composition or a musical segment is actually a complex phenomenon of reception with enormous cultural and personal variation that in most cases cannot be derived from a particular musical constellation.

10.1.5 Machine Learning

In the technique of *supervised learning*, the output of a system is examined in terms of its problem solving ability by an entity that is independent of the system. In a dataset that is used for learning, a single data field may also be explicitly identified as target category. *Reinforcement learning* is an application of supervised learning.

[24] By using the so-called "K-line memory" that reactivates previous behavior when a pattern reoccurs; cf. [40, e.g. p. 60, 84ff].

In this technique, a system receives positive or negative feedback when interacting with its environment, although the system is not provided with an action strategy.

Examples of supervised learning can be found in algorithmic composition in all systems whose outputs are evaluated in regard to particular objectives or by a user[25] and that apply the results as an extension of the knowledge base in order to solve further problems.

In *unsupervised learning*, no target values are given. This method of machine learning aims, among other things, to achieve an autonomous recognition of noticeable patterns in data and to classify them.[26] Unsupervised learning is also applied in data compression which examines the input values in terms of essential components.

Symbol-based machine learning seeks to infer higher orders or general facts from a given amount of data by using mostly inductive methods, whereas problem solving is carried out by sentences formulated in a symbolic language. For the reduction of large state spaces, *inductive biases* that choose efficient procedures on the basis of heuristics are applied. Knowledge may be represented by means of every symbolic representation, such as with first-order predicate logic, object hierarchies, or frames. The concept space makes up the area of all possible solutions that are examined. One possible strategy for the examination is the generalization of statements through, for example, replacing constants by variables or deleting conditions from a conjunctive expression. These generalizations are made on the basis of regularities found in the data material.

Machine learning may be used in supervised or unsupervised form in a number of different procedures. Examples of representative applications in algorithmic composition can be found in chapters 9 and 7. Learning techniques may, however, also make use of statistical procedures, as shown for example by Belinda Thom [59, 60, 61], who designed a software program that, in the framework of unsupervised learning, enables the development of a system for automatic interactive improvisation. In the following, some approaches of symbol-based machine learning are presented which may be divided into rule-based, model-based and case-based techniques.

As examples of rule-based learning, three works will be described briefly, whose knowledge base builds upon rules that are consequently extended by the system. Stephan Schwanauer [54] generates harmonizations by applying different learning models with his system MUSE. On the basis of simple perception models, Gerhard Widmer is developing different problem solving strategies that from correct solutions of musical tasks infer explanations that are then integrated as rules in the learning process. In an exemplary application [66], an upper voice is generated to become a cantus firmus; in an additional work [67], simple melodies are given suggestions for functional harmonic accompaniment.

Schwanauer's system MUSE (Music Understanding System Evolver) fulfils different tasks in the context of four-part harmony. "Learning by rote," "learning from instruction," "learning from failure," and "learning from examples" are the cate-

[25] E.g. Al Biles' system GenJam, cf. chapter 7.

[26] Cf. Kohonen feature maps in chapter 9.

gories within which MUSE provides strategies for the required tasks. Schwanauer also introduces some other interesting approaches, such as "learning by analogy" or "learning from discovery" which, however, are not implemented in his program. In "learning by rote," a number of musical tasks are solved by combining different problem solving entities. The fives stages to be solved are as follows: generation of filling voices for a soprano and bass in root position triads; completing the inner voices for a soprano and figured bass; writing a soprano line and completing the inner voices for a figured bass; writing a soprano line and completing the inner voices for an unfigured bass; harmonizing a chorale melody. As examples of the generation of filling voices, these are selected out of the harmonic material which corresponds to the given soprano and unfigured bass line. At this stage, it is mainly notes in prominent harmonic position that are processed and certain doublings are avoided. A further step examines the results regarding voice leading rules which examine the material for the occurrence of parallel fifths, octaves, and the like. These constraints as absolute prohibitions, together with some directives that do not reject notes outright but that prefer particular notes, make up the exhaustive class of "learning from instruction." If the solutions generated do not correspond to the constraints i.e. in case all attempts to complete the inner voices in the previous steps fail, the state space is once again traversed by means of backtracking, beginning from the last valid solution until a satisfactory result is reached. The results of the backtracking process are stored in MUSE and reflect "learning from failure." If a particular approach proves to be successful in repeating situations, it may be formulated by the program as a new rule – this generalization capability is called by Schwanauer "learning from examples."

In Widmer's [66] system for the generation of upper voices to a given cantus firmus, a musical perception model is established which, together with a training corpus, enables the learning of counterpoint rules. The musical material is organized in a hierarchical system which contains information about key, interval constellations, context of the musical units and the like. The perception model that forms the basis of the system refers to vertical and horizontal interval constellations that are assigned different degrees of harmonic tension. Sequencing events are labeled with properties such as "contrast," "parallelism," or "relaxation." The user evaluates constellations of four notes each with "good," "bad," and "unacceptable." Based on these evaluations, the system inductively generates rules for the musical generation within PROLOG clauses. Figure 10.14 shows a learned generalized rule for the interdiction of "parallel fifths" by a musical configuration classified as "unacceptable."

Furthermore, the system produces hypotheses about possible rules that are then made subject to the user's examination. A determination which may also be applied on the constellation of "parallel fifths" is illustrated in figure 10.15.

These determinations are statements of generalized possibilities that can only be induced to a rule by an exact definition in the framework of a case study. In the above example the general description "bad effect" would have to be concretized by means of a concrete case example through the constellation "perfect consonance" in order to be formulated to a general rule, after positive affirmation by a user. Along

```
unacceptable( Piece, Note) :-
    parallel_note( Note, ParNote),
    previous_note( Note, PrevNote),
    parallel_note( PrevNote, PrevParNote),
    is_interval( Note, ParNote, perfect_consonance),
    is_interval( PrevNote, PrevParNote, perfect_consonance).
```

```
Note = N
ParNote = N1
PrevNote = N2
PrevParNote = N3
```

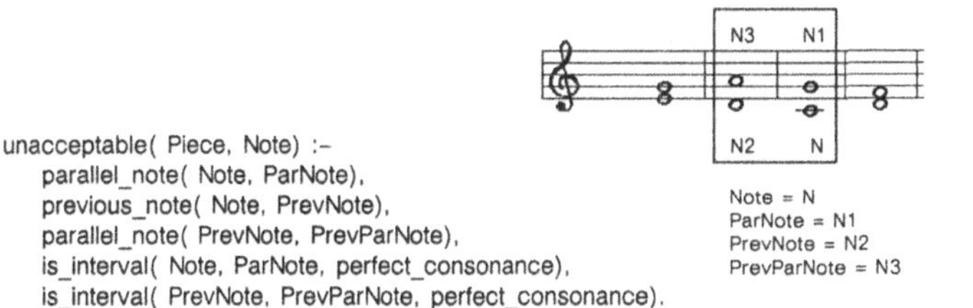

Fig. 10.14 Learned rule from user evaluation [66, p. 495]. © 1992 American Association for Artificial Intelligence.

```
determines( bad_effect( Event1, Effect1, Degree1) &
            bad_effect( Event2, Effect2, Degree2),
          unacceptable_situation( sequence( Event1, Event2))).
```

A sequence *of two events* Event1, Event2 *which are both* bad *may be* unacceptable; *whether or not this is the case depends on exactly what the effects are that make the two events bad* (Effect1, Effect2), *and how strong they are* (Degree1, Degree2).

Fig. 10.15 Determination of possible forbidden constellations. © 1992 American Association for Artificial Intelligence [66, p. 501].

with the determinations, this system also uses heuristics that determine preferences for the order of applied hypotheses.[27]

Widmer's [67] system for the harmonization of simple melodies is based on user evaluations of a training corpus on the basis of which the system generates its harmonization rules. In a hierarchical analysis model with increasing abstraction, the properties of evaluated harmonized songs are related to one another in order to generate criteria for the independent harmonization: "The arrows in Fig. 1 [see figure 10.16] determine what the model can be used for. They describe possible influences between effects; that is, they describe in which way one effect can contribute to the emergence or prominence of another effect. Basically, there are three types of links: positive qualitative proportionalities; negative qualitative proportionalities, and general qualitative dependencies." [67, p. 56]. These dependencies may express themselves in such a way that if, for example, a large distance stands between one chord and the next (along the cycle of fifths) the transition is perceived to be "hard." If, in another case, a note of the melody is part of the chord of the underlying harmony, this area is perceived as harmonically stable. The different forms of musical material, such as tension, relaxation and contrast, are additionally evaluated in a qualitative scale ranging from "low" to "extremely high."

The system generates hypotheses as to why a particular constellation is, for example, perceived to be positive; these are either accepted or denied by the user. Satisfying solutions are integrated into the system as rules for further application. Figure 10.17 shows an extract of an explanation for a melodic-harmonic constellation that is evaluated as positive.

[27] For a more detailed description of determinations and heuristics, see [66, p. 498ff].

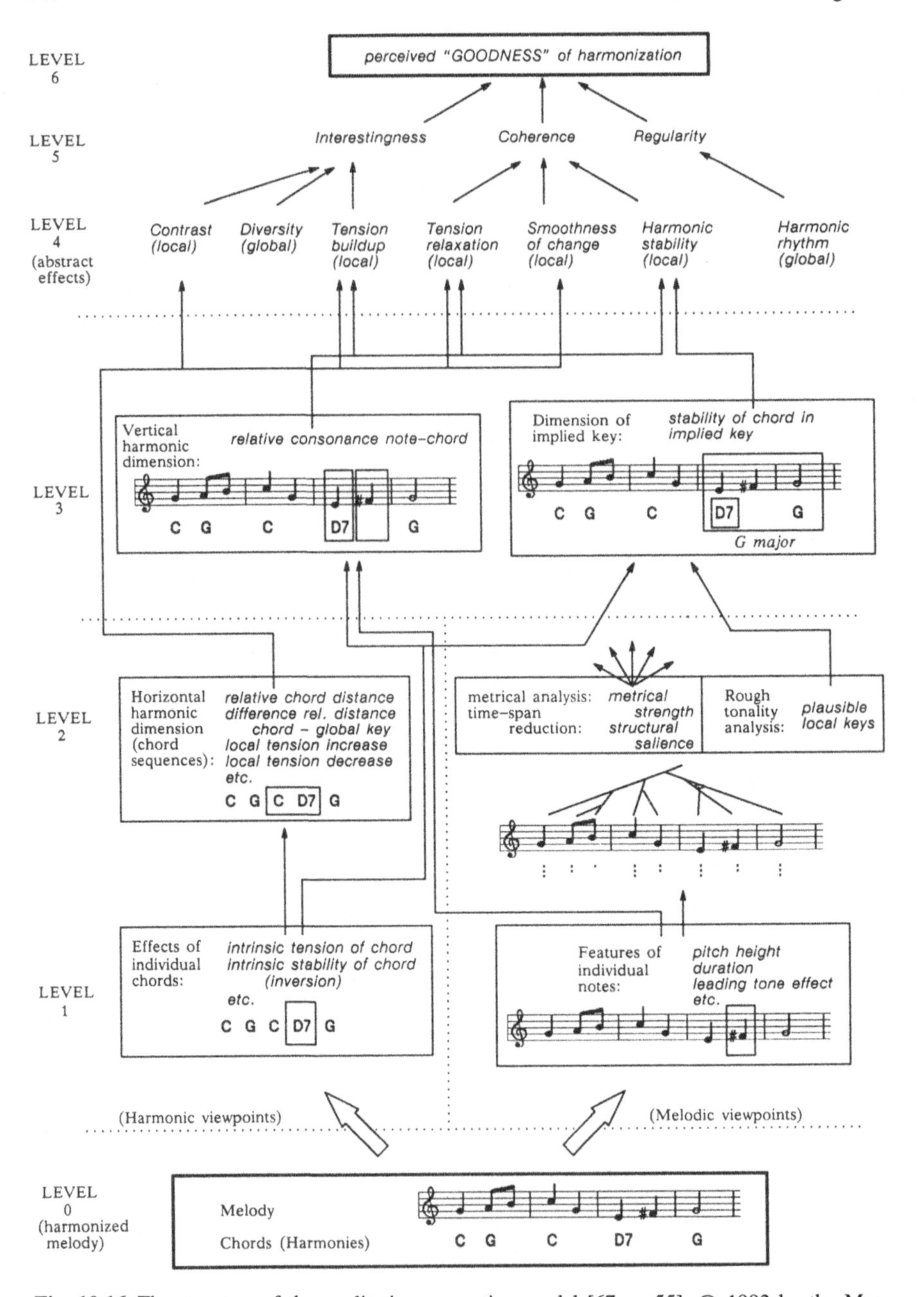

Fig. 10.16 The structure of the qualitative perception model [67, p. 55]. © 1992 by the Massachusetts Institute of Technology.

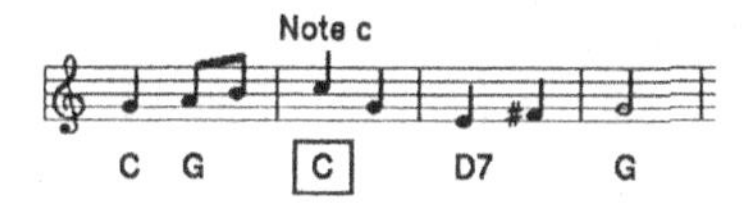


```
;;; Top 2 levels of explanation (see model in Fig.1)

(6) good( C, c)  because                                       ;;; C is a good harmonization for note c because
(5) |   coherence( C, c, extremely_high) because                ;;; C creates a sense of coherence because
    |   |   previous-chord( C, G)                               ;;; chord C is preceded by chord G
(4) |   and tension-relaxation( C, c, moderate)                 ;;; it moderately relaxes some existing tension
(4) |   and smoothness-of-change( G, C, extremely_high)         ;;; and the change from G to C is smooth
(4) L   L and harmonic-stability( C, c, extremely_high)         ;;; and the chord is highly harmonically stable


;;; Structure of complete explanation

(6) good( C, c)  because
(5) | coherence( C, c, extremely_high)  because
    | |   previous-chord( C, G)                                   ;;; chord C is preceded by chord G in the piece
(4) | | and tension-relaxation( C, c, moderate)  because
(2) | | |   local_tension_decrease( G, C, moderate)  because
    | | |   |   global-key( c_major)                              ;;; there is some moderate relaxation of tension
(1) | | |   and distance_from_tonic(G,c_major,1)                  ;;; because the C chord is closer to the global
(1) | | |   and distance_from_tonic(C,c_major,0)                  ;;; tonic than its predecessor G
    | | |   L and 1 > 0
(2) | | | and salience( c, high)  because                         ;;; the relaxation effect is perceived even more
    | | |   |   metrical_strength(c, high)                        ;;; strongly as the point where it occurs (note c)
    | | L   L and structural_salience( c, high)                   ;;; is of high metrical and structural salience
                                                                  ;;; (determined from time-span reduction)
(4) | | and smoothness-of-change( G, C, extremely_high)  because  ;;; the transition from G to C is
(2) | | |   relative_chord_distance( G, C, low)  because          ;;; smooth because G and C are
    | | L   L   distance_on_circle_of_fifths( G, C, 1)            ;;; just one perfect fifth apart

(4) | | and harmonic-stability( C, c, extr._high)  because        ;;; C is highly harmonically stable because
(3) | | |   relative_consonance( C, c, extr._high)  because       ;;; 1.: the note c is consonant with
    | | |   |   scale_contains_note( c_major, c)                  ;;;     the scale implied by the chord
    | | |   L and chord_contains_note( C, c)                      ;;;     and with the chord itself

(2) | | | and plausible_local_key( c, c_major)                    ;;; 2.: C is stable within the
(3) | | | and stability_of_chord_in_key(C,c_major,extr.high)  because   locally implied key (c major)
(1) | | |   L   tonic-chord( C, c_major)                          ;;;     because it is the tonic chord
(1) | | | and intrinsic_stability_of_chord( C, high)  because     ;;; 3.: and in addition, as a
    | | |   L   chord_type( C, triad)                             ;;;     root position triad, the
    L L L                                                         ;;;     C chord is intrinsically stable
```

Fig. 10.17 Explanation in Widmer's system for a satisfying melodic-harmonic constellation [67, p. 58]. © 1992 by the Massachusetts Institute of Technology.

A *model-based system* analyzes a problem domain by means of functions and specifications of relevant reference models. Errors can be recognized through the discrepancy between expected and observed behavior. One application of a technique designed in a similar way can be found in the work of Rens Bod [8], who analyzes examples of folk music. Based on the principles of "Musikalische Gestalt,"[28] compositions of a corpus are compared in terms of similarities in the structural form of segments.

Case-based reasoning systems (CBR) make their decisions on the basis of already gained experience and usually run through a four-step process when solving a problem:

1. Cases whose solutions may also be applied to solving the target problem are retrieved from the memory.
2. The solution of the case that best matches the target problem is used for an initial problem solving.
3. After mapping the previous solution to the target case, it is revised and adapted to the target situation.
4. After the adaptation, the new case and the resulting experience are stored in the memory for use in further cases. The system is therefore able to learn and extends its problem solving competence with every new case.

Several works will now be introduced briefly as examples of the application of CBR in the context of algorithmic composition.

Francisco Pereira, Carlos Grilo, Luìs Macedo and Amilcar Cardoso [43] developed a software system which, based on a corpus of some compositions of the Baroque age, generates soprano voices after the input of a bass voice. In the analysis, the musical pieces are divided according to a hierarchical tree-like structure whose nodes represent the musical terminals. Similar to the principles of a generative grammar, the material is subdivided into phrases and sub-phrases, etc. up to the single notes. The different nodes are given additional indications for position, meter, tonality, and previous and succeeding nodes as well. The system may also comprehend musical structures that are completely modified by functions such as transposition. Each of these musical objects is formulated within Prolog clauses and represents the case memory of the CBR system. When a new bass voice is given as an input, the system searches cases in the divided case database and generates a soprano voice based on them, preferring solutions that differ more from the corpus. In order to use the state space more efficiently, backtracking is applied.

A similar approach is being pursued by Paulo Ribeiro, Francisco Pereira, Miguel Ferrand and Amilcar Cardoso with their system MusaCazUza [49]. Here, too, a soprano voice is produced, which, however, has a harmonic line as an input. The case

[28] Developed by Christian Ehrenfels (1859–1932), founder of the "Gestalt theory" that states in simple words that perception of phenomena happens as a complex whole and not as the sum of its constituent parts. Ehrenfels demonstrates musical Gestalt e.g. by means of the transposability of melody progressions whose concrete note values change, although the melody may still be recognized.

database of this system consists of harmonic elements (scale degrees) in combination with rhythmically structured melodic phrases. The context-dependent attributes of each case are, for example, pitch and duration of the notes building up the melody, the duration of the harmonic function, the harmonic function itself, information on the context regarding melody and harmony and the amplitude and direction between the first and the last notes in the case. After the input of a harmonic progression, the system generates a sequence of possible cases for the production of the soprano voice. Several possible cases are arranged by means of different criteria; the best variant in each case is integrated in the output sequence. After obtaining a melody, the user may alter the results and integrate them as a new case into the database.

The generations of these systems are, in some cases, described by the authors to be satisfying; however, disadvantages result from the application of a comparatively small corpus – a limiting fact that is also explicitly mentioned in the work of Pereira et al. [43, p. 5].

Jordi Sabater, Joseph L. Arcos and Ramon López de Màntaras [53] are attempting to solve this problem in their CBR system for the harmonization of melodies by involving rule-based techniques. A limitation of the state space in the case database is reached in this system through the indication of "representative notes" that correspond mainly to longer notes in the first half of the measure. Together with the underlying harmonies, these notes constitute the cases of the database. If no appropriate case can be found for an input, the rule-based component of the system is applied in which some simple principles for the generation of harmonic progressions are encoded. The rule-based solutions are added to the CBR database and are available in the case memory in case of a new generation of the system.

The system SaxEx by Arcos, Sabater and de Màntaras generates agogic and dynamic structure by means of a CBR system [3, 4, 5]. SaxEx uses spectral analysis to extract information on parameters such as articulation, dynamics, rubato and vibrato of a saxophone interpretation and produces a database on the basis of the performance data. As an input for the system, an "inexpressive" interpretation may be used that is arranged by SaxEx with agogic and dynamic variations. Another possibility is to use a musically already interpreted performance and derive the agogic and dynamic information for the expressive interpretation of other pieces of music. The case data base in SaxEx consists of musical structures that are arranged according to the principles of Lerdahl and Jackendoff's generative theory of tonal music. Results of the CBR system that may be modified by the user are available as new cases in the database. In an extended version of SaxEx, musical movements are assigned differing shares of basic affective dimensions that may also be further applied explicitly for the creation of the performance. A three-dimensional space serves to represent musical affections. Each of these axes in that space represents a scale between two contrasting affections, such as "tender – aggressive," "sad – joyful," and "calm – restless." Then, a position is assigned on each of these axes to the musical segment; therefore, for the whole emotional aspect of the musical segment a clear position in this representation space results. Although the assignment of affective qualities to musical structure is, in principle, a difficult undertaking, this representation form by Arcos et al. is, however, an interesting alternative for a differentiated "affective" labeling of musical aspects.

10.1.6 Agents

In contrast to a central knowledge base, agents are models of cooperating intelligence in the sense of a distributed problem solving approach. The agent does not require knowledge about the superior or parallel problem solving entity in order to complete its task; it acts autonomously. To differentiate such a construct from the concept of sub-programs or simple objects, some necessary properties that the program must possess in order to be considered an agent are given. Luger and Stubblefield define an agent as "an element of a society that can perceive (often limited) aspects of its environment and affect that environment either directly or through cooperation with other agents [35, p. 16]. The authors give a number of criteria for its characterization: An agent is "autonomous" to a certain degree which means that it performs some tasks of problem solving to the most part independent of other agents. Further, it is "situated" in an own surrounding and has a certain knowledge of its environment. An agent is "interactional," referring to it being part of a collection of individual agents that it collaborates with to solve tasks. Regarding the society of all agents, they coordinate in order to find a solution. Consequently the phenomenon of "emergence" occurs, meaning that through the cooperation of the collective of agents a result is reached which has more significance than the contribution of each individual agent [35, p. 15–16].

Lämmel's [33] definition of an agent even includes several additional properties, for example: Agents are "continuous" (acting continuously for some time), "mobile" (e.g. movement in a network), and also have "personality" that may guarantee a particular appearance of the agent. All these criteria have something in common: They imagine an agent to possess "autonomy"; a property that, similar to the term "intelligence," may be interpreted in various ways.

Mikhail Malt [37] developed a multi-agent system that generates musical output in the context of an artistic project related to the improvisation of an instrumentalist. Each agent in Malt's system possesses a name, a fixed spatial position, a central note, a perception radius (a region of listening), a task (a musical gesture), a maximal life span, and a behavior that is expressed in different interactions with circumjacent agents. The musical gesture of each agent may be an element of a melody, a sound sample, or something similar. Within a particular perception radius, interactions with other agents occur that appear in two basic behavioral patterns: In "social" behavior, the musical parameters of adjacent agents assimilate, whereas in "anti-social" behavior these parameters develop in contrary directions to their neighbors. "Social" and "anti-social" behaviors are defined in this system by the relation of the root tone of an agent to the average value of the root tones of other agents within the perception radius. In order to enable the population size of the system to remain the same despite the limited life span of the single agents, "reproduction" may be applied by transferring particular parameters such as name, position, or task of an agent to a "descendant." The interrelations between all agents produce the musical output of the system: "As the action of each agent is always dependent on the other agents, the notion of emergence was fundamental in this experiment. This notion expresses the appearance of a new meaning during the aggregation of elements within a given context." [37, p. 7].

Joseph Rukshan Fonseka [20] developed an agent type that interacts with other agents by communicating in a computer network. The starting point for Fonseka's work is a description for the generation of musical material of the British composer Cornelius Cardew: "Each chorus member chooses his own note (silently) for the first line [...] For each subsequent line choose a note that you can hear being sung by a colleague. [...] Time may be taken over the choice. If there is no note, or only the note you have just been singing, or only 2 notes or notes that you are unable to sing, choose your note for the next line freely. Do not sing the same note on two consecutive lines. [...] Remain stationary for the duration of a line; move around only between lines." [20, p. 27–28]. The description given here in a shortened version makes up the seventh section of Cardew's experimental musical work "The Great Learning," written in 1967. Music is herein described using instructions, for example, in which tones are equated with states and circumjacent choir singers with the environment; consequently, this concept shows a conceptual parallel with the principles of cellular automata. In the realization of this concept in a network, Fonseka uses the latency of data transmission between networked computers as an analogy to the physical distance of the choir singers.

The fact that agent types that originally had not been developed for musical tasks[29] may also be applied in algorithmic compositions in interesting ways is shown by Tatsuo Unemi and Daniel Bisig [64] in the context of an interactive installation which employs algorithms to generate music and visuals. In this work, the movements of a user affect the activities of agents moving in a virtual three-dimensional space – the graphical and musical output results from the interplay of real and virtual movement. Each agent has two "instruments" whose panorama, tone pitch and velocity (values on the x, y and z axes) are acquired from the present spatial position of the agent. Apart from rules that determine the possible movements of each agent within the flock, the user may influence the selection of the instrument as well as the positioning of the entire flock. In order to limit the number of musical events, different strategies are applied that allow only certain agents to produce tones. User interfaces enable the definition of a tone pitch range, the selection of particular scales or also the indication of chord progressions within whose pitches the agents may contain their musical activities.

[29] Here: *flocking agents*; agents that may show complex patterns of movement resembling flocks of fish or birds in their graphic representation. The first application was developed by Craig Reynolds with "BOIDS," a multi-agent system for the simulation of flocks of birds; cf. [48]. The work of Unemi and Bisig is mentioned only as an example, similar approaches can also be found e.g. in [51] [57] and [7]; for an overview, see [64]. Further interesting reading should be found in the newly published book edited by Miranda and Biles [41].

In most of the multi-agent systems in the context of algorithmic composition, the output of the system proves to be an emergent phenomenon – a complex, often unpredictable musical result that is produced by the interaction of mostly simply structured subprograms (the agents). This behavior, which is normally difficult to foresee, can also be found in the conceptually related cellular automata as well as in some works in the field of genetic algorithms.[30] But it is exactly this aspect of complexity and unpredictability that may represent an interesting approach: "From these experiments it is possible to put forth the hypothesis that a musical surface could be seen as a system to which an unstable dynamic will correspond, driven by a multiplicity of forces in interaction. The composition will then be seen as a process in permanent movement, a permanent search for meaning between the different levels of the considered musical space, with moments of stabilization, moments of destabilization and mainly the phenomena of emergence."[31]

10.2 Synopsis

Due to the numerous approaches of classification, the representation of algorithmic composition in the framework of artificial intelligence is quite difficult. The classification made in this chapter leaves some works that are treated in the context of other techniques unmentioned. So it would indeed be possible to represent, for example, generative grammars or Markov models as tools of language processing within the scope of AI. The large quantities of literature available for methods like these, however, are in this work a motivation to treat them separately. For the procedures subsumed in this chapter under the concept of AI, the following concluding reflections may be made.

The peculiarity of musical information implies, at least in the generation of style imitations, enormous state spaces. Brute-force techniques fail to handle the complexity and diversity of the musical material. The basic problem here is the fact that the musical components that are to be processed can only be connected to larger units in a meaningful way through a complex approach. Even when, for example, only one parameter such as pitch is treated, the result is related to problems such as the correct handling of the context and the often necessary recognition of a hierarchical structure that is implicitly available in the corpus. Furthermore, when a larger number of parameters are considered and polyphonic structures are modeled, this increases the amount of possible outputs considerably. For these reasons, systems of algorithmic composition in the framework of AI often apply heuristic strategies for limiting the state space; however, heuristics cannot naturally guarantee optimal solutions. In this regard, variants of context-related knowledge representation that limit the musical event space to expected configurations must also be mentioned. For the consideration of a larger number of parameters, *multiple-viewpoint systems*

[30] Exemplary of that is [12], see chapter 7.

[31] Malt on the results of his work; cf. [37, p. 7].

that may represent and process the musical material in terms of different aspects are used frequently in AI.

In algorithmic composition, methods of logical reasoning and their extensions are, above all, used for the representation and derivation of knowledge – such as, for example, in the form of predicate calculus – or also for the limitation of parameter areas by fuzzy sets. Non-monotonic reasoning is not an explicit procedure of algorithmic composition; however, due to the fact that, in a creative process, musical premises cannot have the validity of axioms in the sense of a "musical truth," non-monotonic logic represents a good analogy to a compositional process during which musical premises are, in most cases, taken as a basis and may indeed be revised in the further course of the creation of the work.

The evaluations of the outputs of algorithmic composition systems in the framework of AI are subject to the same restrictions as a fitness evaluation in a genetic algorithm. Apart from the already mentioned disadvantages of a user evaluation, effective rule-based evaluations are not easy to perform in light of the complexity of musical information. Methods of machine learning enable an improved approach by allowing the induction of rules from a corpus in a self-acting way. Even if a newly learned rule must also often be examined by a user with respect to its generalization capability, with every new rule the adaptive system extends its problem solving capacity. Because learning happens in these systems exclusively by means of examples, techniques of machine learning therefore adapt optimally to the specific situations in a given corpus. Similar advantages are also provided by methods of case-based reasoning that, in the generation of new material, examine a corpus in terms of "similar" solutions without having to possess domain-specific knowledge.

Aside from style imitation tasks, the field of multi-agent systems provides a large number of interesting possibilities for algorithmic composition. The phenomenon of emergence produces a complex system behavior that is much more than the total of its individual parts – or expressed differently in analogy to the thinking of Douglas R. Hofstadter: "Every aspect of thinking can be viewed as a high-level description of a system which, on a low level, is governed by simple, even formal rules." [27, p. 559].

References

1. Agon C, Assayag G, Bresson J (eds) The OM composer's book, 1. Delatour France, Sampzon; Ircam, Paris ISBN 2-7521-0028-0
2. Ames C, Domino M (1992) Cybernetic Composer: an Overview. In: Balaban M, Ebcioglu K, Laske O (eds) Understanding music with AI. AAAI Press/MIT Press, Cambridge, Mass. ISBN 0262-52170-9
3. Arcos JL, Lopez de Mantaras R, Serra X (1997) SaxEx: A case-based reasoning system for generating expressive musical performances. In: Proceedings of the 1997 International Computer Music Conference. International Computer Music Association, San Francisco
4. Arcos JL, Canamero D, Lopez de Mantaras R (1998) Affect-driven generation of expressive musical performances. Emotional and intelligent: The tangled knot of cognition, papers from

the 1998 AAAI Fall Symposium. AAAI Technical Report FS 98-03. AAAI Press, Menlo Park, Calif
5. Arcos JL, Lopez de Mantaras R (2002) AI and music. Case based reasoning. From composition to expressive performance. AI Magazine, 23/3, 2002
6. Balaban M, Ebcioglu K, Laske O (eds) Understanding music with AI. AAAI Press/MIT Press, Cambridge, Mass. ISBN 0262-52170-9
7. Blackwell TM, Bentley PJ (2002) Improvised music with swarms. In: Proceedings of the 2002 Congress on Evolutionary Computation, vol 2. IEEE, New York, pp 1462–1467
8. Bod R (2002) Probabilistic grammars for music. In: Proceedings BNAICSaxEx: A case-based reasoning system for generating expressive musical performances 2001, Amsterdam
9. Casey MA (2003) Musical structure and content repurposing with bayesian models. In: Proceedings of the Cambridge Music Processing Colloquium, University of Cambridge, 2003
10. Cope D (2001) Virtual music: computer synthesis of musical style. MIT Press, Cambridge, Mass. ISBN 0-262-03283-X
11. Courtot F (1992) Logical representation and induction for computer assisted composition. In: Balaban M, Ebcioglu K, Laske O (eds) Understanding Music with AI. AAAI Press, California. ISBN 0262-52170-9
12. Dahlstedt P (1999) Living Melodies. http://www.design.chalmers.se/projects/art_and_interactivity/living-melodies/ Cited 26 Mar 2005
13. Dannenberg R (1984) An on-line algorithm for real-time accompaniment. In: Proceedings of the 1984 International Computer Music Conference. International Computer Music Association, San Francisco
14. Ebcioglu K (1992) An expert system for harmonizing four-part chorales. In: Schwanauer SM, Levitt DA (eds) Machine models of music. MIT Press, Cambridge, Mass. ISBN 0-262-19319-1
15. Elsea P (1995) Fuzzy logic and musical decisions. http://arts.ucsc.edu/ems/music/research/FuzzyLogicTutor/FuzzyTut.html Cited 1 Jan 2005
16. Encyclopedia Britannica Online (2006) http://www.britannica.com/dictionary?book=Dictionary&va=heuristic&query=heuristic Cited 17 Jan 2006
17. Feigenbaum EA, Barr A (1981) The handbook of artificial intelligence, I. Pitman, London. ISBN 0-273-08540-9
18. Feigenbaum EA, Barr A (1981) The handbook of artificial intelligence, II. Pitman, London. ISBN 0-273-085553-0
19. Feigenbaum EA, Cohen PR (1981) The handbook of artificial intelligence, III. Pitman, London. ISBN 0-273-08554-9
20. Fonseka JR (2000) Musical agents. Thesis, Electrical and Computer Systems Engineering Department of Monash University, 2000
21. Friberg A, Colombo V, Frydén L, Sundberg J (2000) Generating musical performances with Director Musices. Computer Music Journal, 24/3, 2000
22. Fry C (1984) Flavors Band: A language for specifying musical style. Computer Music Journal, 8/4, 1984
23. Görz G (1993) Einführung in die künstliche Intelligenz. Addison-Wesley, Bonn. ISBN 3 89319 507 6
24. Henz M, Lauer S, Zimmermann D (1996) CompoZe intention-based music composition through constraint programming. In: Proceedings of the 8th International Conference on Tools with Artificial Intelligence (ICTAI 96). IEEE Computer Society, Los Alamitos, Calif, pp 118–121
25. Hiller L, Isaacson L (1993) Musical composition with a high-speed digital computer. In: Schwanauer SM, Levitt DA (eds) Machine models of music. MIT Press, Cambridge, Mass. ISBN 0-262-19319-1
26. Hiller L (1963) Informationstheorie und Computermusik. Darmstädter Beiträge zur neuen Musik, 8. Schott, Mainz

27. Hofstadter D (1979) Goedel, Escher, Bach: An eternal golden braid. Basic Books, New York. ISBN 0465026850
28. Jacob BL (1995) Composing with genetic algorithms. In: Proceedings of the 1995 International Computer Music Conference. International Computer Music Association, San Francisco
29. Jacob BL (1996) Algorithmic composition as a model of creativity. Organised Sound, 1/3, Dec 1996
30. Johnson ML (1992) An expert system for the articulation of Bach fugue melodies. In: Baggi D (eds) (1992) Readings in computer-generated music, IEE Computer Society Press, Los Alamitos, Calif. ISBN 0-8186-2747-6
31. Katayose H, Inokuchi S (1989) The Kansei Music System. Computer Music Journal, 13/4, 1989
32. Kreutzer W, McKenzie B (1991) Programming for Artificial Intelligence. Methods, tools and applications. Addison-Wesley, Sydney. ISBN 0 201 41621 2
33. Lämmel U, Cleve J (2001) Lehr- und Übungsbuch Künstliche Intelligenz. Fachbuchverlag Leipzig im Carl-Hanser-Verlag, München. ISBN 3-446-21421-6
34. Levitt DA (1993) A representation of musical dialects. In: Schwanauer SM, Levitt DA (eds) Machine models of music. MIT Press, Cambridge, Mass. ISBN 0-262-19319-1
35. Luger GF, Stubblefield W (1998) Artificial intelligence. Structures and strategies for complex problem solving, 3rd edn. Addison Wesley Longman, Harlow. ISBN 0-805-31196-3
36. Homepage of The Loebner Prize in Artificial Intelligence (2007) http://www.loebner.net/Prizef/loebner-prize.html Cited 2 May 2007
37. Malt M (2001) In Vitro – Growing an artificial musical society. http://galileo.cincom.unical.it/esg/Music/workshop/articoli/malt.pdf Cited 2 Jun 2005
38. Manzolli JA, Moroni F, Von Zuben R, Gudwin R (1999) An evolutionary approach applied to algorithmic composition. In: Proceedings of SBC'99 – XIX National Congress of the Computation Brazilian Society, Rio de Janeiro, 3, 1999
39. Mazzola G (2002) The topos of music. Geometric logic of concepts, theory, and performance. Birkhäuser, Basel. ISBN 3-7643-5731-2
40. Minsky M (1988) The society of mind. Simon & Schuster, New York. ISBN 0-671-60740-5
41. Miranda ER, Biles JA (eds) (2007) Evolutionary computer music. Springer, London. ISBN 978-1-84628-599-8
42. Pachet F (2001) Musical harmonization with constraints: A survey. Constraints, 6/1, 2001
43. Pereira FC, Grilo C, Macedo L, Cardoso A (1997) Composing music with case-based reasoning. In: Proceedings of the Second Conference on Computational Models of Creative Cognition, MIND-II, Dublin, Ireland, 1997
44. Phon-Amnuaisuk S, Smaill A, Wiggins G (2002) A computational model for chorale harmonisation in the style of J.S. Bach. In: Proceedings of ICMPC7 (the 7th International Conference on Music Perception and Cognition), Sydney, Australia, 2002
45. Phon-Amnuaisuk S (2004) Logical representation of musical concepts (for analysis and composition tasks using computers). In: Proceedings of SMC'04 (Sound and Music Computing), Paris, France
46. Ponce de Leòn PC, Pérez-Sancho C, Iñesta JM (2004) A shallow description framework for musical style recognition. In: Fred A, Caelli T, Duin RPW, Camphilo A, de Ridder D (eds) Structural, syntactic, and statistical pattern recognition: joint IAPR international workshops SSPR 2004 and SPR 2004, Lisboa, Portugal, August 18–20, 2004; procedings. Lecture notes in computer science, vol 3138. Springer, Berlin, pp 876–884
47. Ramirez R, Peralta J (1998) A constraint-based melody harmonizer. In: Proceedings ECAI'98 Workshop on Constraints for Artistic Applications, Brighton
48. Reynolds C (1987) Flocks, herds and schools: A distributed behavioural model. Computer Graphics, 21/4, 1987, pages 25-34
49. Ribeiro P, Pereira FC, Ferrand M, Cardoso A (2001) Case-based melody generation with MuzaCazUza. In: Proceedings of the AISB'01 Symposium on Artificial Intelligence and Creativity in Arts and Science, 2001

50. Riecken DR (1992) Wolfgang: A system using emotional potentials to manage musical design. In: Balaban M, Ebcioglu K, Laske O (eds) Understanding music with AI. AAAI Press/MIT Press, Cambridge, Mass. ISBN 0262-52170-9
51. Rowe R (1992) Machine listening and composing with Cypher. Computer Music Journal 16/1, 1992
52. Russell S, Norvig P (1995) Artificial intelligence. A modern approach. Prentice Hall, Englewood Cliffs, NJ. ISBN 0-13-103805-2
53. Sabater J, Arcos JL, de Màntaras LR (1998) Using rules to support case-based reasoning for harmonizing melodies. Multimodal reasoning: papers from the 1998 AAAI Spring Symposium. AAAI Technical Report SS-98-04. AAAI Press, Menlo Park, Calif
54. Schwanauer SM (1993) A learning machine for tonal composition. In: Schwanauer SM, Levitt DA (eds) Machine models of music. MIT Press, Cambridge, Mass. ISBN 0-262-19319-1
55. Shieber SM (1994) Lessons from a restricted Turing test. Communications of the Association for Computing Machinery, 37/6, 1994
56. Spangler RR (1999) Rule-based analysis and generation of music. Thesis, California Institute of Technology, Pasadena, California
57. Spector L, Klein J (2002) Complex adaptive music systems in the breve simulation environment. In: Bilotta E et al (eds) ALife VIII: workshop proceedings, Sydney, NSW, pp 17–23
58. Taube H (2004) Notes from the metalevel: An introduction to algorithmic music composition. Routledge, London. ISBN 9026519753
59. Thom B (1999) Learning melodic models for interactive melodic improvisation. In: Proceedings of the 1999 International Computer Music Conference. International Computer Music Association, San Francisco
60. Thom B (2000) Artificial Intelligence and real-time interactive improvisation. In: Proceedings of the Seventeenth Conference on Artificial Intelligence (AAAI-2000), Workshop on Artificial Intelligence and Music, Austin, Texas
61. Thom B (2000) Unsupervised learning and interactive Jazz/Blues improvisation. In: Proceedings of the Seventeenth Conference on Artificial Intelligence. AAAI Press/MIT Press, Cambridge, Mass, pp 652–657
62. Turing AM (1950) Computing machinery and intelligence. Mind 59, 1950
63. Wechsler D (1964) Die Messung der Intelligenz Erwachsener. Textband zum Hamburg Wechsler Test für Erwachsene (HAWIE). Hans Huber, Bern
64. Unemi T, Bisig D (2004) Playing music by conducting BOID agents. In: Pollack J, Bedau MA, Husbands P, Ikegami T, Watson RA (eds) Artificial life IX: proceedings of the Ninth International Conference on the simulation and synthesis of living systems. MIT Press, Cambridge, Mass pp 546–550
65. Weizenbaum J (1966) ELIZA – A computer program for the study of natural language communication between man and machine. Communications of the ACM, 9/1, January 1966
66. Widmer G (1992) The importance of basic musical knowledge. A knowledge intensive approach to machine learning. In: Balaban M, Ebcioglu K, Laske O (eds) Understanding music with AI. AAAI Press, Cambridge Mass. ISBN 0262-52170-9
67. Widmer G (1992) Qualitative perception modeling and intelligent musical learning. Computer Music Journal 16/2, 1992
68. Winston PH (1993) Artificial Intelligence, 3rd edn. Addison-Wesley, Reading, Mass. ISBN 0 201 53377 4

Chapter 11
Final Synopsis

Procedures of algorithmic composition may be used for the treatment of single aspects of a musical task, or for determining the overall structure of a musical piece. These two alternatives imply the use of very specific approaches and aesthetic positions regarding the generation of a composition. Therefore, creating a conceptual division may further clarify the different application possibilities of particular algorithmic techniques. Algorithmic composition is used either for creating genuine, original compositions, or in the field of style imitation, where musical material is generated according to a given style or represents an attempt to verify a musical analysis by resynthesis. The fact that the procedures described in this work are mainly used for tasks of style imitation, may partly be explained by the environment the authors are working in, and their motivation for applying algorithms to the generation of musical structure.

11.1 Algorithmic composition as a genuine method of composition

If procedures of algorithmic composition are exclusively used for the realization of an original compositional strategy, then one may speak of algorithmic composition as a genuine method of composition. In this case, one must bear in mind that "genuine methods of composition" cannot be defined precisely, since there may also exist style imitations of a proprietary "style" by, for example, using algorithmic methods to realize one specific aesthetic concept in several different works. Furthermore, the integration of common algorithmic procedures of musical structure generation, such as canonic or serial techniques, may also be seen as style imitation on a structural level. Other gray areas in definition between genuine composition and style imitation may result due to the cultural setting: Genuine composition may occur in a specific cultural environment, but as soon as some of its aspects are reconstructed by means of resynthesis, e.g. in the field of music ethnology, one would sooner regard it as style imitation. Depending on the period of its creation, a historical approach may classify a musical piece generated by some set of rules as a "genuine composition," or as a style imitation. Finally, this aspect is inherent in most procedures of algorithmic composition through the generation of a framework for the production of a whole class of compositions – consequently, each concrete generation is also a style-compliant production of a general underlying aesthetic concept. These possible gray areas should only make clear that the terms "genuine composition" and "style imitation" cannot be neatly separated. To simplify this, a dividing line could be drawn between the two approaches in regard to the motivation for applying algorithmic techniques: Hence, algorithmic composition could be referred to as a "method of composition" if procedures are applied in the context of the creation of a new musical piece of art. Accordingly, "style imitation" is the attempt to model a style that is established in musicology, from a historic or ethnologic perspective, as a particular genre. Herein, not all rules of the genre under examination must necessarily be encompassed, since the modeling may also be carried out using non-knowledge-based methods.

If techniques of algorithmic composition are used as "genuine composition methods" in the sense of this definition, the question arises as to what extent algorithmic principles contribute to the generation of a composition. This aspect of algorithmic composition ranges from the complete determination of a structure to accompanying procedures of a compositional process. The first case refers to "algorithmic composition" in the strictest sense. Although the total structure of a musical piece is modeled here, not all possible musical parameters need to be algorithmically determined in every case. In most computer music systems of algorithmic composition, the parameters pitch, duration, and dynamics may be manipulated; any further differentiation in the articulation of details of musical events, like particular instrumental playing techniques, is generally not performed. This may well be explained by the continuing dominance of the MIDI protocol developed in the 1980s, and still used in many systems as an interface today, which by default only allows for the mapping of the three abovementioned parameters. "Computer Assisted Composition" (CAC), in a narrower sense of the expression, uses algorithmic procedures as supplementary compositional tools. In the literature, the terms "algorithmic composition" and "Computer Assisted Composition" are often used synonymously. Whether the application of algorithmic principles is carried out with or without the aid of a computer is of secondary importance in the end – a fact that is also borne out in numerous historical applications of rule-based composing (see chapter 2).

Algorithmic composition may also be examined for the extent to which algorithmic principles are applied in the generation of a composition. If algorithmic procedures are used in the process of composition as additional tools, they may either be used to structure formal relationships or to generate musical material itself. In this context, Xenakis, for example, uses Markov models to control the sequential order of musical sections. Schönberg's twelve-tone technique provides him with material which is then further modified in the course of the process of composition. Algorithmic models, regardless of their use for the formal structuring or the generation of new material, may also only represent compositional possibilities; the generations serve as an inspiration here, without necessarily using the results of the algorithm as concrete musical material in the final work. What these approaches have in common is that composition is understood to be a discursive process, which in its course is influenced by individual preferences. This approach to algorithmic composition finds its parallel in the non-monotonic logic of AI. Here, the algorithms formulated for the compositional concept do not have the validity of axioms – on the contrary, the results of the generative process may need to be critically examined during the creative process. So in this case, the algorithmic procedures become a compilation of useful tools that are used deliberately for the generation of a composition. Here, a large part of the creative work is not determined by algorithms – individual human intervention and original ideas remain essential aspects of this approach.

An alternative possibility when using algorithmic principles lies in the decided intention to put the algorithm in the position of the composer. The composition becomes an expression of a transpersonal principle, as can be seen in Hauer's work, or questions the traditional social role of composer, interpreter and recipient, as with Cage. In addition, mapping strategies that aim at establishing "validity" of compositional structure by trying to map systems valid for specific extra-musical fields on musical structures may be subsumed under this category.

Furthermore, procedures of algorithmic composition may be applied for solving combinatorial tasks. In this context, the creation of a system that continually produces new musical material is often a strong incentive for the application of algorithmic procedures. The musical dice game (see chapter 2), which in the course of history was likely generated by various composers, is a good example for this category. In this approach, the motivation to meet highest quality standards of composition is not the primary focus.

Algorithmic principles may also be inherent in a musical genre. The isorhythmic motet, counterpoint, and different types of canons are only a few forms that are composed within the framework of a complex rule system. Although composing in these genres is traditionally not understood to be algorithmic composition, the rules are at least formulated as constraints or restricting conditions, which is one reason for the frequent use of these and similar forms for applications of style imitation.

Procedures of algorithmic composition may also be used for implementing aesthetic principles in order to examine their musical applicability. A prime example of this approach is the "Illiac Suite" (see chapter 10) by Hiller and Isaacson, who, in their "experiments," generated musical structures within generate-and-test cycles by means of Markov models and other stochastic principles and constraints.

Another possible approach is the application of algorithmic procedures as a composition language, determining the overall structure of a piece while performing modifications solely on the algorithmic level, but not altering the output. This method should not be mixed up with a conventional generate-and-test cycle. Here, stochastic alternatives are used to generate a class of compositions out of which musical pieces are then selected according to individual preferences. Creative manipulation of the algorithm takes the place of personal selection – the algorithm is further modified and finally becomes able to generate a very specific composition. In this approach, the creation of the algorithm leads to treating the generations of a system from a creative perspective – the algorithm, turned deterministic, replaces personal selection and becomes able to represent a precisely formalized aesthetic concept of a single possible composition.

Moreover, methods of algorithmic composition exist that primarily serve commercial interests. A form of the musical game of dice, consisting of arbitrarily combinable playing cards, was marketed as a parlor game in early 20th century Boston.[1] Although the sales did not meet the high expectations, the game marked the beginning of the commercial marketing of a product which enables algorithmic composition. Software systems such as BandInABox[2] were not actually developed with a fundamentally musical ethos in mind, but were predominantly focused on a broad public as a marketable product. Algorithmically generated music may – far from any claim to meet compositional criteria of quality – itself become the subject of primarily commercial interests: Functional music for department store chains, waiting loops for telephone systems, ring tones for mobile phones and the like, are increasingly attractive branches of business. Software solutions here churn out pleasing "compositions" as if off an assembly line; no musical education is needed for the operation of systems of this kind – and above all, there are no royalties to be paid.

11.2 The dominance of style imitation in algorithmic composition

The fact that procedures of algorithmic composition are, despite the wide variety of possible applications, mainly used in the field of style imitation, may be partly explained by the background of the authors who place their works in the public domain. The majority of the works mentioned have been developed in the fields of human and natural sciences, in which publishing is considered part of a scholar's everyday work. In musicology, style imitation is mostly intended to verify analysis approaches by resynthesis. In this sense, a comprehensive modeling of all possible musical parameters is not intended or even feasible due to the complexity of the examined style.[3] Works that are developed in a natural scientific environment often focus on an examination of the informatic properties of the applied algorithm; here, musical information only represents one possible form of data. The fact that the employed musical styles have been examined by musicologists in terms of an explicitly applied or implicit rule system seemingly allows an easy-to-perform evaluation of the generated material.

On the other hand, publications on algorithmic techniques for genuine composition are rare – a fact that may be to a certain degree explained by the professional environment and the aesthetic positions of composers. Moreover, the evaluation of compositional results is mainly subject to subjective preferences. Provided that com-

[1] The so-called Kaleidacousticon System, cf. [11, p. 823].

[2] Software that produces a desired spectrum of musical pieces of different musical provenience on the basis of chord progressions.

[3] Cf. the problems of generative grammar in music-ethnological research and Blacking's objections in chapter 4.

posers are university educated, however, the study of algorithmic techniques – although not explicitly formulated as such – represents an essential part of their education. Subjects such as counterpoint and harmony are disciplines that are to a high degree determined by rules, and practicing them teaches students to work in formalizable systems. A number of musicological concepts complement this approach with analytical methods. So, even during education, composition is taught as a field of art which can be formalized to a certain degree – a fact that implies the application of algorithmic procedures in future compositional work. Composers publishing material on their algorithmic methods is rather a rare occurrence and may be explained to a certain extent due to the following consideration: Publishing is in general not part of the profession. It is replaced by a composition, whose creative principles – if formalized – are not willingly communicated for obvious reasons. In addition, there are often basic objections to the use of algorithmic principles, because the musical structure becomes comprehensible and the composition therefore cannot claim "ingenious creativity" any longer. Associated with that is the fear of "disenchantment" at a piece's creative merit, which may also explain why algorithmic composition, alongside the abovementioned reasons is mostly only used as a supplementary tool in genuine composition. In consideration of this fact, it is gladly accepted that the influence of algorithmic composition on the overall structure of the musical piece remains marginal; essential decisions are left to the "creative" ideas of the composer. If, however, interesting musical structures are generated by procedures of algorithmic composition, these frequently mentioned objections hold less ground as arguments. Even though the musical result is completely determined by the underlying formal principles, thus becoming seemingly banal, the creative process has, in actual fact, only used a different language. Composition no longer takes place in the notation of concrete note values, but in the realization of a compositional concept by means of specially created formalizable processes. In both cases the musical result is what counts – its "notation" is a question of the individual approach.

Composers are often also not sufficiently familiar with informatic methods to be able to implement their structural ideas within a programming language. Although algorithmic composition is not principally bound to the computer, more demanding concepts are, however, hardly feasible with "pencil and paper." Naturally, there exists a wide range of programs that at the push of a button also map complex algorithms on musical parameters; however, the possibilities for influencing the modeling of the composition are limited mostly to the selection of particular scales or rhythmic values used for the mapping. Accordingly, the results are arbitrary or again represent simple style imitations that may be generated through the selection of categories such as "Techno," "New-Age," or "Experimental."

When writing about the results of algorithms for "genuine composition," contrary to the output of "push and play" programs, they are difficult to judge, because here no reference corpus or preceding analysis is available. In this context, algorithmic composition is finally also a question of individual approach and musical objectives.

So, considered from the point of view of publishability, style imitation will be the main field of application for algorithmic composition. However, a number of innovations and further developments in highly specialized computer music environments enabling algorithmic composition to almost any degree of complexity, also reveal a great interest in developing a creative approach towards the generation of musical structure, even though the pieces produced by these systems are in most cases not published as a structural analysis of the applied algorithms.

11.3 Origins and characteristics of the treated procedures

Virtually all procedures used for tasks of algorithmic composition have their origins in extra-musical fields, and often become highly popular outside the purely academic study of these disciplines. Chaos theory, cellular automata, Lindenmayer systems, artificial intelligence, but also neural networks, are subjects of a broad non-scientific discussion and are sometimes represented as comprehensive procedures that claim to be solutions of universal validity. In this sense, for example, even the title of Wolfram's book on cellular automata, "A New Kind of Science," suggests a revolution in the scientific landscape.[4] Similarly, even by 1987, James Gleick was predicting a paradigm shift in physics with "Chaos: Making a New Science."[4]. In the 1980s, chaos theory became extremely popular due to the wide adoption of some aspects of the works of Edward N. Lorenz and Benoit Mandelbrot – the so-called "butterfly effect,"[5] self-similarity, and the visually fascinating illustrations of the Mandelbrot set are all phenomena that helped chaos theory to attract extraordinary public attention. Lindenmayer systems appear in their graphic representation as intriguing plant-like structures. Neural networks present a completely new approach to sub-symbolic information processing – the fact that these systems are based on a biological model further increases public interest. In the field of artificial intelligence, programs such as ELIZA make the vision of thinking machines often seem real; consequently, AI becomes a prevalent part of the repertoire of Science Fiction. With Stanley Kubrick's "2001: A Space Odyssey" in 1968, if not earlier, Hollywood turned artificial intelligence into a deadly threat. In Kubrick's film, it is the computer HAL 9000 which – actually a friendly and caring companion – exterminates a spaceship crew due to a conflict in his programming. In Andy and Larry Wachowski's 1999 movie "The Matrix," intelligent computer programs appear as "agents," keeping mankind in ignorance and slavery.

The general popularity of some procedures explains their application in algorithmic composition as well. Cellular automata and neural networks are, despite the fact that they were developed for applications other than musical tasks, often used in the generation of musical structure. Mapping the behavior of chaotic systems may yield interesting musical results that are, however, subject to a number of restrictions in terms of intervention and structuring possibilities. In addition to the techniques of

[4] [14]; in this context, also see Wolfram's statement in chapter 8.

[5] Small modifications performed on the initial conditions of equations for a weather model developed by Lorenz, lead to strongly varying results. This behavior is illustrated with the example of a butterfly which could by one flap of its wings (inducing turbulence) influence the meteorological conditions at a distant location.

chaos theory, Lindenmayer systems are also well suited to the realization of self-similar concepts in the generation of musical structure. Furthermore, in algorithmic composition, the heterogeneous research field of artificial intelligence is unable to provide universally valid solutions; rather it is a collection of different methods that may be applied for the representation and processing of musical information.

The applicability of procedures of algorithmic composition may also be considered from the following perspective: In which context and for what application was the algorithm originally developed? The background of a method may explain difficulties that are above all related to the peculiar character of musical information. For example, Lindenmayer systems were developed for the simulation of the growth process of plants. In the modeling of a fern, the L-system shows, in the graphic representation of the iteration cycles, increasingly complex branching, until eventually the "natural" density of the plant structure has been reached. The set of symbols of an L-system is in general very limited. Therefore, when these symbols are mapped on note values without carrying out further measures, a large number of similar events are produced. In order to achieve a larger variability in the generated material, the mapping strategy must be modified in different ways[6] and the increase of the systems must be treated musically.[7]

Similar problems may also be found in other procedures that were originally not developed for musical disciplines. Generative grammar originating in the field of linguistics is in principle well suited to the processing of one-dimensional symbol strings; this results in the fact that mutual horizontal and vertical dependencies of musical information can only be processed when the original formalism is extended.[8] Neural networks, on the other hand, were originally developed for image processing and classification; when they are applied in algorithmic composition, it is above all the treatment of the temporal context that requires the modification of network types or the search for an appropriate representational form for the musical data.[9]

Different methods may also be distinguished due to their specific suitability for tasks of algorithmic composition. Procedures such as generative grammars may be applied both in genuine composition and the field of style imitation. Special cases are cellular automata and Markov models, which are either only used for genuine composition (CA), or almost exclusively[10] in the field of style imitation (MM). An essential criterion for the suitability of a particular approach is given by the possibilities of encoding the musical information. In many cases, it is already these greatly differing ways of encoding that determine the possible applications of a particular procedure. In this sense, generative grammars may, due to their rewriting rules, for-

---

[6] Cf. e.g. DuBois' strategies in chapter 6.

[7] Interesting issues regarding the musical realization of self-similar structures can also be found in a theoretical work by composer Bernhard Lang: "Diminuendo. Über selbstähnliche Verkleinerungen" [5].

[8] Cf. software Bol Processor by Bernhard Bel in chapter 4.

[9] Cf. Mozer's system CONCERT in chapter 9.

[10] Here, also, exceptions prove the rule, cf. the applications of MM for tasks in genuine composition as performed by Xenakis and Hiller.

mulate good descriptions of a given musical structure. This similarly applies to a Markov model, which is nearly exclusively created on the basis of an underlying corpus. The rules of a cellular automaton, on the other hand, are hardly able to describe a particular corpus, but are the starting point for the changing of cell states. Of course, procedures may be modified to a considerable degree and applied in an originally unintended way – but, to quote a Buddhist analogy: Does it make sense to pave a street with leather instead of putting on shoes?

The way in which a specific procedure outputs the data, also plays an essential part. A generative grammar produces its terminals only at the end of all substitutions – this is a singular process which in its temporal steps may not be traced back musically, since non-terminals that remain in the symbol strings up to the last derivation do not yet represent concrete musical information. A genetic algorithm, however, produces generations of material until it is stopped or the fitness criteria have been fulfilled. A cellular automaton continuously modifies the states of its cells during its cyclic sequence, whereas in contrast to the genetic algorithm, no temporal limit is given by the fulfillment of a particular objective. This difference in behavior alone brings about distinct fields of application – here, there are principally two concepts that can be distinguished: The first generates single outputs as the result of a calculation process or a parsing; the other produces a continuous data flow which may be understood as a musical process in time, be it target-oriented or not. Naturally, a generative grammar may enable a permanent musical output by means of a special application using stochastic rewriting rules, and a genetic algorithm may only be made audible in its last generation – interesting studies, however, often refer to the peculiarities of the output of a particular procedure.[11]

11.4 Strategies of encoding, representation and musical mapping

An algorithm that is to represent musical information in an appropriate way requires some sort of data encoding for its inputs, as well as specific strategies for musical "mapping," meaning the way the outputs of the algorithm are mapped on musical parameters. These aspects of algorithmic composition are of crucial importance since they present the interfaces between information processing and the musical structure. Both the encoding and the representation provide musical information for processing in an algorithm. However, a comparison of these procedures[12] illustrates two distinct principles that are nevertheless closely related: Encoding is a necessary precondition for the processing of a priori existing musical information in an algorithm; representation, on the other hand, maps the output of the algorithm on

[11] Cf. the "Living Melodies" of Dahlstedt and Nordahl in chapter 7 or Dorin's Boolean Sequencer in chapter 8.

[12] "Encode: to change information into a form that can be processed by a computer," "Representation: the act of presenting sb/sth in a particular way," both definitions from [7]. Frequently, the term "representation" is used for "encoding," e.g. "An individual member of a population is represented as a string of symbols." [2].

note values in different ways, or displays the musical material with regard to one or more different aspects. Degrees of a tempered scale, for example, may be encoded as binary or decimal values; in both variants of encoding, the musical material may be represented in an absolute way – here, a particular number corresponds to a particular note value. Also, for example, distances in a tonal space may be encoded in binary or decimally for a relative representation – in this case, the information on the concrete note values may only be obtained through the preceding context. Representation may take place within local or distributed models. In a local scheme, for example, a particular note value may be assigned to a grid column of a cellular automaton, or a certain neuron on a neural network. In a distributed encoding, on the other hand, e.g. the activation states of distinct neurons of a particular layer provide the information for determining a note value. Although the representation of musical information does not modify the specific behavior of an algorithm, it has strong influence on the form of the output since it also presents a musical interpretation of the produced data.

While during encoding musical information is prepared as optimally as possible for its processing through an algorithm, representation is responsible for providing the encoding with the type of information which best characterizes the musical material. The effects of absolute (e.g. by concrete note values) representations are easy to manage and changes only happen locally – any modification of the context may be disregarded. In this sense, in an absolute representation, the mutation of a genetic algorithm only changes the corresponding note value and the succeeding values remain untouched in regard to their musical interpretation. A relative (e.g. intervallic distances of a melodic movement) representation enables a generalized illustration of pitches by abstracting away from concrete keys; however, by changing a value, all succeeding values are also modified – a property of relative representation that may also lead to undesired consequences. Through this representation, for example, the result of the mutation in a genetic algorithm becomes overly complicated, and in neural networks, a net error has an affect on all succeeding values. Multi-dimensional forms of representation correspond to the potential complexity of the underlying musical material and enable the display of musical parameters under different aspects. Mozer, for example, uses three representational forms for the parameter 'pitch' in his system CONCERT (see chapter 9).Another interesting approach of this kind is developed by Cope in his EMI system, which enables the illustration of a musical component on different hierarchical levels.[13]

In the mapping of the outputs of the algorithm on musical parameters, the applied strategy has great influence on the structure of the generated material. Outputs that are restricted within particular limits may well be mapped on musical parameter ranges by means of scaling. In algorithms that, for example, produce monotonously ascending or descending values, modulo operations may be used to limit the outputs to appropriate ranges. However, this method may also cloud the specific behavior of an algorithm, if it does not consider, for example, the toroidal structure of a cellular automaton. In case a musical mapping requires a scaling of the data, this can be

[13] By means of the classification module SPEAC, see chapter 5.

achieved, for example, by a rhythmical quantization, or by means of an assignment which maps value ranges on particular musical events. Quantization and assignment may, however, also make the results of the algorithm and its musical outputs arbitrary. In case a rhythmic structure is quantized too roughly, this may cause a stylistically relevant parameter to become unrecognizable. But also, if complex outputs of an algorithm are mapped on only a few pitches which are "melodious" in all combinations, a pleasing sound texture is generated, though the structure-forming algorithm becomes exchangeable.

Such arbitrary mapping strategies raise the question of whether the idiosyncrasies of an algorithm should always be reflected in the musical output as a matter of principle. Basically, regardless of the underlying algorithm, in every case the form of musical output will be of primary significance; if, however, the characteristics of the applied algorithm are not considered in either way within the mapping strategy, the motivation for its application must be questioned. If the idiosyncrasies of an algorithm are taken into account in the mapping process, this fact does not mean that the underlying structure is or should be perceptually comprehensible as well. Furthermore, the complexity of an algorithm is of lesser significance to the musical logic. Strategies, whose musical effects are easily predictable, such as the application of rewriting rules or methods like serialism, allow for interesting musical results. Disregarding the intentional generation of unpredictable material for "entering new musical territory," these "simple" concepts may be used to effectively put into practice decidedly intended musical objectives. Frequently used mapping strategies apply the results of already performed mappings in non-musical domains as a starting point, and consequently represent a 'mapping of a mapping.' Examples of this can be found in the interpretation of graphical representations of algorithms, such as in the musical mapping of turtle graphics of a Lindenmayer system or in the association of columns of a CA-grid with particular pitch information. In principle, there are no theoretical objections to this form of mapping; however, the respective strategies may make essential properties of an algorithm unrecognizable: The self-similarity of an LS is no longer apparent in the musical structure; the cells of a CA-grid become, contrary to their usual function, switches of values, that are only dependent on their particular position.

In algorithmic composition, the parameters 'pitch' and 'duration' are used mostly for the mapping. Systems like SaxEx, by Arcos et al., dealing as they do with the modeling of agogic and dynamic interpretation, are exceptions. Attempts which go beyond that and try to analyze emotional qualities in musical information, or to apply them as a means of a compositional process, reduce these perceptive phenomena to easily comprehensible structural properties and therefore often remain arbitrary in regard to their strategies.[14]

Algorithmic composition aims to either generate style-compliant material or to realize a compositional principle. From these approaches, the technique of sonification must be distinguished, which makes particular properties of an underlying data material auditorily available. Making the results of sonification aesthetically

[14] For these works, see chapter 10.

appealing – suggesting a close relationship to algorithmic composition as a genuine artistic discipline – is, however, not a necessary precondition. The mapping, which consequently may also be musically motivated, here takes a completely different approach and serves a fundamentally different purpose.

11.5 The evaluation of generated material

The evaluation of generated material is an essential aspect in the application of procedures of algorithmic composition. Two basic approaches may be distinguished, which either examine the suitability of an algorithm for particular data structures or evaluate the musical output. In the first category are, for example, comparisons that examine algorithms for their suitability for treating a musical context.

The evaluation of an output may either be performed by a user, or algorithmically – here, the aforementioned problems of the fitness function arise. The alternative user evaluation is often difficult to objectify, since in many cases subjective or vague statements, such as those which follow, are made on the results of musical generations: "After sufficient training, GenJam's playing can be characterized as competent with some nice moments." [1, p. 6]; "Most musical pieces created sounded very reasonable." [10, p. 7]; "SICOM compositions are comparable to a young student's with the first degree of Analysis and Composition." [9, p. 6]. In "Towards a Framework for the Evaluation of Machine Compositions," [8] Marcus Pearce and Geraint Wiggins therefore claim a scientifically verifiable evaluation of the outputs of algorithmic composition systems. They differentiate between the evaluation strategies of the system, which finally lead to an output ("critic"), and the final evaluation of the generated material ("evaluation"). Different alternatives are described regarding the second case. In addition to the abovementioned subjective rating carried out by a human, the situation in a concert as well as an algorithmic evaluation are also considered problematic: An algorithmic rating often only implies a subjective opinion; in a concert, which should through the audience balance the subjectivity of an individual opinion, the differing levels of knowledge, as well as the different individual musical preferences, may be difficult to manage in terms of a quality criterion in the field of style imitation. As to original composition, this form of evaluation is considered to be a suitable method: "However, while a well received performance would seem a good criterion for the evaluation of new works (as in the case of Biles) [...]" [8, p. 3]. Examining this argument, one may object that Biles's system also acts in the context of an established style and therefore could be regarded as style imitation. In addition, the taste of the audience may indeed be a measure for the acceptance of a composition, but hardly for the quality of a piece of art. On the basis of Turing's test, Pearce and Wiggins decided to use an evaluation performed by human users in a number of tests series, with the task being to distinguish between machine-generated and "human composed" musical fragments.[15]

[15] Cope subjects the outputs of EMI to a musical Turing test as well, see chapter 10.

In this work, the attempt to establish an objective evaluation sets a clear focus on the examination of the output of algorithmic composition systems – an important aspect, which, however, is neglected in many approaches in this field. Instead, in most cases a detailed investigation is only carried out in regard to the architecture of the system; here, the output is a pleasing by-product and often merely a confirmation of the functionality of the system. Mozer, too, explicitly points to this weakness in regard to the generation of musical structure by means of neural networks: "One potential pitfall in the research area of connectionist music composition is the uncritical acceptance of a network's performance. It is absolutely essential that a network be evaluated according to some objective criterion. One cannot judge the enterprise to be a success simply because the network is creating novel output." [6, p. 195].

11.6 Limits of algorithmic composition

In general, procedures of algorithmic composition may be divided into knowledge-based and non-knowledge-based methods. Knowledge-based approaches generate their outputs often on the basis of a rule-based system which is formulated by if-then conditions and/or constraints. Non-knowledge-based methods are able to autonomously derive rules from an underlying corpus and produce outputs that, in supervised learning, are additionally evaluated by a superior instance. Both systems are well suited to both genuine composition and the generation of style imitations.

The processing of knowledge within a rule-based system does not pose any problems for genuine composition; in analogy to the Closed World Assumption[16] of AI, the following argument may be applied to this field: The rules, which may be extended arbitrarily, present a creative instrument of artistic interpretation and reflect in any case the intentions of the user – that which has not been modeled is also irrelevant here. The rules of a knowledge-based system which apply to style imitation may, however, come into conflict with the other implicit rules of the style to be modeled – though the algorithm may conform to the Closed World Assumption, for the desired style it would require all of its criteria to be without exception comprehended by the rule system. This objective is, in most cases, not feasible. In general, according to the terminology of generative grammar, the highest generative capacity possible is aimed for: The task is to model all aspects of a style and at the same time exclude all incorrect constructs from the generation. First, high generative capacity brings along the problems of high complexity, and second, due to the peculiarity of the algorithm, correct (in the sense of well-formed) structures may be produced that are, however, not allowed in the respective style. This means that, for example, incorrect movements may be produced by a generative grammar, since the local positioning of a musical movement contradicts a musical rule that has not been acquired. Although this problem may be theoretically met with a very restrictive set of rules, as a consequence, the output of the system is reduced.

[16] Everything that cannot be proved explicitly is assumed to be false, this meaning that everything that is not modeled is also irrelevant to the respective model.

In general, the rules and constraints of a knowledge-based system disregard the generation of a single musical structure, and describe a class of possible compositions instead. The applicability of a number of possible rules, and the basic conditions formulated by the constraints, create a stochastic scope which enables the abstraction from a single case. Within these alternative possibilities, a wide variety of solutions are possible; but, it is not guaranteed that each of these generations is appropriately generated in terms of the style to be modeled, and even if the criterion of well-formedness should be fulfilled, this is not a sufficient condition for an aesthetically suitable solution. This stochastic scope, which may include a number of solutions, is opposed to a highly restrictive algorithmic description that, in the worst case, only reproduces examples of the corpus.

For genuine composition – as one possible approach – the finally determined output of an algorithm (after a repeated modification) does not present a problem; here, a new composition is produced as a result of dealing creatively with the algorithm.

The possible applications of non-knowledge-based systems are limited by the structure of the corpus and the aligned possibilities of the necessary derivation of information from data. In applications of style imitation, the data mostly consists of pieces of a particular genre. Systems such as neural networks, or methods of machine learning, generalize information available in a musical corpus in order to generate a class of "similar" compositions. In general, optimal procedures do not exist here, since the generalization capability of a procedure strongly depends on the type of data material, as well as on the form of the desired output. If the generation of smaller musical segments is the objective, genetic algorithms or neural networks may be well applied for this task. These methods prove to be less suited for the modeling of larger structures. In this case, procedures like grammatical inference, which generate hierarchical rewriting rules on the basis of a corpus, can be used – a restriction exists insofar as only state transitions can be encoded in the rules that also exist in the corpus.

Furthermore, the structure of the corpus is to a large extent jointly responsible for the quality of the generated output. The selected examples should be able to sufficiently represent the class of possible compositions and at the same time, depending on the strategy of the algorithm, these musical pieces should be able to be combined with each other – this means that the corpus must not join several classes of compositions whose structural properties are not interchangeable. Generally, the previously mentioned problems are only relevant in the context of style imitation, as within genuine composition, algorithms may be optionally formulated according to a concept of composition – here, a conflict between the algorithmic program and the musical result can only occur through an insufficient formalization of the individual compositional requirements.

In most cases, systems of algorithmic composition generate – analogous to the score in occidental music – a symbolic level. However, since beyond this level music is a complex phenomenon of interpretation and reception, this fact points out another limitation of generating musical structure using algorithmic procedures.

11.7 Transpersonalization and systems of "universal" validity

Algorithmic composition is often motivated by the desire to transpersonalize the process of composition and to establish an unchallengeable quality criterion through the referencing of the musical structure to its however "scientifically" generated basis. In all these attempts, circular reasoning is inherent: An idea is elevated to an axiom on whose basis a model of composition of unchallengeable validity may be asserted. As soon as a "primary principle" of musical creation is formulated, the quality of musical structure becomes explainable due to this principle; compositions meeting the highest quality criteria will virtually produce themselves through the application of rules that have once been accepted as valid. In such a philosophical approach – not regarding musical quality – it is of secondary importance whether we are talking about a "Zwölftonspiel" in Hauer's system or a composition based on Schillinger's[17] rules. A similar principle can already be found in the establishment of the truth of the "Ars Magna," the difference in this case being that the axioms are asserted from Christian dogmatism (see chapter 2). Regardless of the quality of the generated musical structure, the problem of such theoretical concepts is apparent and produces justifiable doubts in regard to algorithmically generated music. A possible way to avoid this evident dilemma may be to undertake the abovementioned discursive examination of musical structure on an algorithmic level: In this context, applying the algorithm here only represents a rather unusual compositional method, whereas neither axiomatic validity nor transpersonal significance are inherent in this approach, just as is the case in every other creative approach bringing into being a piece of art.

11.8 Concluding remark

"To some extent, this match is a defense of the whole human race. Computers play such a huge role in society. They are everywhere. But there is a frontier that they must not cross. They must not cross into the area of human creativity. It would threaten the existence of human control in such areas as art, literature and music."[18] This statement, made by Garry Kasparov, who in 1997 as Chess World Champion was defeated by the supercomputer Deep Blue in a tournament, illustrates reservations and fears that arise when considering automated "creativity." If programs play chess on the level of grand masters, this does not appear so strange; but when a computer invades a creative artistic field, this is met with skepticism. However – algorithmic composition is not a musical golem, usurping creativity from the human

[17] Schillinger was also a well-known teacher of theory. His composition system, cf. [12] became popular e.g. through the "Moonlight Serenade," created as an exercise by his student Glenn Miller. The following citation illustrates Schillinger's evaluation of compositional systems: "The final step in the evolution of the arts is the scientific method of art production, whereby works of art are manufactured and distributed according to definite specifications." [13, p. 6].

[18] Garry Kasparov, cited after [3, p. 40].

realm. The algorithm is a tool and means for the creative examination of the complex aspects of musical production. Or with the words of HAL 9000: "I've got the greatest enthusiasm and confidence in the mission."

References

1. Biles JA (1994) GenJam: A genetic algorithm for generating Jazz solos. In: Proceedings of the 1994 International Computer Music Conference. International Computer Music Association, San Francisco
2. Biles JA (1995) GenJam Populi: Training an IGA via audience-mediated performance. In: Proceedings of the 1995 International Computer Music Conference. International Computer Music Association, San Francisco
3. Cope D (2001) Virtual music: Computer synthesis of musical style. MIT Press, Cambridge, Mass. ISBN 0-262-03283-X
4. Gleick J (1987) Chaos: Making a new science. Penguin Books, New York. ISBN 0-14-00 9250-1
5. Lang B (1996) Diminuendo. Über selbstähnliche Verkleinerungen. Beiträge zur Elektronischen Musik, 7. Institut für Elektronische Musik (IEM) an der Universität für Musik und darstellende Kunst in Graz, Graz
6. Mozer M C (1991) Connectionist music composition based on melodic, stylistic, and psychophysical constraints. In: Todd PM, Loy DG (eds) Music and connectionism. MIT Press, Cambridge, Mass. ISBN 0-262-20081-3
7. Oxford Advanced Learner's Dictionary (2006) http://www.oup.com/oald-bin/web_getald7index1a.pl. Cited 17 Jan 2006
8. Pearce M, Wiggins G (2001) Towards a framework for the evaluation of machine compositions. In: Proceedings of the AISB 2001, Symposium on AI and Creativity in the Arts and Sciences
9. Pereira FC, Grilo C, Macedo L, Cardoso A (1997) Composing music with case-based reasoning. In: Proceedings of the Second Conference on Computational Models of Creative Cognition, MIND-II, Dublin
10. Pigg P (2002) Cohesive music generation with genetic algorithms. http://web.umr.edu/ tauritzd/courses/cs401/fs2002/project/Pigg.pdf. Cited 11 Nov 2004
11. Roads C (1996) The computer music tutorial. MIT Press, Cambridge, Mass. ISBN 0-262-68082-3
12. Schillinger J (1973) Schillinger System of musical composition. Da Capo Press, New York. ISBN 0306775522
13. Schillinger J (1976) The mathematical basis of the arts. Kluwer Academic/Plenum Publishers, New York. ISBN 0306707810
14. Wolfram S (2002) A new kind of science. Wolfram Media, Champaign, Ill. ISBN 1-57955-008-8

Index